Blackwell's Concise Encyclopedia of ECOLOGY

Contents

List of Contributors

P.A. **P. ALBERCH** [Deceased] *Museo Nacional de Cienceas Naturales, J. Gutierrez Abascal 2, Madrid E-28006, Spain*

P.M.S.A. **P.M.S. ASHTON** *School of Forestry and Environmental Studies, Yale University, 360 Prospect Street, New Haven, CT 06511, USA*

J.A. **J. AUSTIN** *UK Meteorological Office, London Road, Bracknell RG12 2SZ, UK*

D.D.B. **D.D. BALDOCCHI** *Atmospheric Turbulence and Diffusion Division, US Department of Commerce, 456 South Illinois Avenue, PO Box 2456, Oak Ridge, TN 37831, USA*

J.S.B. **J.S. BALE** *School of Biological Sciences, University of Birmingham, Edgbaston, Birmingham B15 2TT, UK*

C.J.B. **C.J. BARNARD** *School of Biological Sciences, University of Nottingham, University Park, Nottingham NG7 2RD, UK*

R.S.K.B. **R.S.K. BARNES** *Department of Zoology, University of Cambridge, Downing Street, Cambridge CB2 3EJ, UK*

R.W.B. **R.W. BATTARBEE** *Environmental Change Research Centre, University College London, University of London, 26 Bedford Way, London WC1H 0AP, UK*

B.B. **B. BAUR** *Department of Integrative Biology, Section of Conservation Biology (NLU), University of Basel, St Johanns-Vorstadt 10, CH-4056 Basel, Switzerland*

J.B. **J. BEDDINGTON** *T.H. Huxley School of Environment, Earth Sciences and Engineering, Royal School of Mines, Prince Consort Road, London SW7 2BP, UK*

M.B. **M. BERAN** *OB Research Services, 1 The Croft, East Hagbourne, Didcot OX11 9LS, UK*

G.P.B. **G.P. BERLYN** *School of Forestry and Environmental Studies, Yale University, 370 Prospect Street, New Haven, CT 06511, USA*

D.A.B. **D.A. BERRIGAN** *Department of Zoology, Box 351800, University of Washington, Seattle, WA 98195, USA*

M.C.M.B. **M.C.M. BEVERIDGE** *Institute of Aquaculture, University of Stirling, Stirling FK9 4LA, UK*

S.J.B. **S.J. BISSELL** *Colorado Division of Wildlife, Denver, CO 80216, USA*

L.O.B. **L.O. BJÖRN** *Section of Plant Physiology, Lund University, Box 117, 221 00 Lund, Sweden*

J.F.B. **J.F. BLYTH** *Institute of Ecology and Resource Management, University of Edinburgh, Mayfield Road, Edinburgh EH9 3JU, UK*

C.B. **C. BOESCH** *Max-Planck Institute for Evolutionary Anthropology, Inselstrasse 22, 04301 Leipzig, Germany*

P.B. **P. BORRELL** *EUROTRACISS, GSF-Forschungs zentrum für Umwelt und Gesundheit, Kühbachstrasse 11, D-81543 München, Germany*

P.M.B. **P.M. BRAKEFIELD** *Institute of Evolutionary and Ecological Sciences, University of Leiden, PO Box 9516, NL-2300 RA Leiden, The Netherlands*

D.R.B. **D.R. BROOKS** *Department of Zoology, University of Toronto, Toronto, Ontario M5S 3G5, Canada*

M.A.B. **M.A. BURGMAN** *School of Forestry, University of Melbourne, Creswick, Victoria 3363, Australia*

G.L.B. **G.L. BUSH** *Department of Zoology, Michigan State University, East Lansing, MI 48824, USA*

J.C. **J. CAIRNS JR** *Department of Biology, Virginia Polytechnic Institute and State University, Blacksburg, VA 24061, USA*

P.C. **P. CALOW** *Department of Animal and Plant Sciences, University of Sheffield, Sheffield S10 2TN, UK*

J.N.C. J.N. CAPE *Institute of Terrestrial Ecology, Bush Estate, Penicuik EH26 0QB, UK*

T.C. T. CARACO *Department of Biological Sciences, State University of New York at Albany, Albany, NY 12222, USA*

H.C. H. CASWELL *Biology Department, Woods Hole Oceanographic Institution, Woods Hole, MA 02543, USA*

R.C. R. CATCHPOLE *English Nature, Over Haddon, Derbyshire DE45 1JE, UK*

W.G.C. W.G. CHALONER *Department of Geology, Royal Holloway, University of London, Egham Hill, Egham TW20 0EX, UK*

C.W.C. C.W. CLARK *Department of Mathematics, University of British Columbia, Vancouver, British Columbia V6T 1Z2, Canada*

D.C. D. CLODE *Department of Zoology, University of Melbourne, Parkville, Victoria 3052, Australia*

S.G.C. S.G. COMPTON *Ecology and Evolution Research Group, School of Biology, University of Leeds, Leeds LS2 9JT, UK*

L.M.C. L.M. COOK *The Manchester Museum, University of Manchester, Oxford Road, Manchester M13 9PL, UK*

R.M.C. R.M. COWLING *Institute for Plant Conservation, University of Cape Town, Rondebosch 7701, South Africa*

R.M.M.C. R.M.M. CRAWFORD *Plant Sciences Laboratory, Sir Harold Mitchell Building, University of St Andrews, St Andrews KY16 9AL, UK*

M.J.C. M.J. CRAWLEY *Department of Biology, Imperial College of Science, Technology and Medicine, Silwood Park, Ascot SL5 7PY, UK*

J.Cr. J. CRESSWELL *Department of Biological Sciences, University of Exeter, Hatherby Laboratories, Exeter EX4 4PS, UK*

R.H.C. R.H. CROZIER *Department of Genetics and Human Variation, La Trobe University, Bundoora Campus, Bundoora, Victoria 3083, Australia*

P.J.C. P.J. CURRAN *Department of Geography, University of Southampton, Highfield, Southampton SO17 1BJ, UK*

K.D. K. DAVIES *SAC, Crops Division, Bush Estate, Penicuik EH26 0PH, UK*

R.W.D. R.W. DAVIES *Faculty of Science, Monash University, Wellington Road, Clayton, Victoria 3168, Australia*

A.J.D. A.J. DAVIS *Ecology and Evolution Research Group, School of Biology, University of Leeds, Leeds LS2 9JT, UK*

J.C.D. J.C. DEUTSCH *Crusade, 73 Collier Street, London N1 9BE, UK*

D.A.D. D.A. DEWSBURY *Department of Psychology, University of Florida, Gainesville, FL 32611, USA*

M.Di. M. DICKE *Department of Entomology, Wageningen Agricultural University, PO Box 8031, NL-6700 EH Wageningen, The Netherlands*

M.D. M. DOEBELI *Zoology Institute, University of Basel, Rheinsprung 9, CH-4051 Basel, Switzerland*

G.A.D. G.A. DOVER *Department of Genetics, University of Leicester, University Road, Leicester LE1 7RH, UK*

C.D. C. DYTHAM *Department of Biology, University of York, PO Box 373, Heslington, York YO1 5YW, UK*

D.E. D. EBERT *Zoology Institute, University of Basel, Rheinsprung 9, CH-4051 Basel, Switzerland*

J.E. J. EKMAN *Department of Zoology, Uppsala University, S-752 36 Uppsala, Sweden*

M.A.E. M.A. ELGAR *Department of Zoology, University of Melbourne, Parkville, Victoria 3052, Australia*

J.R.E. J.R. ETHERINGTON *Parc-y-Bont, Llanhowell, Solva, Haverfordwest SA62 6XX, UK*

D.A.F. D.A. FALK *Society for Ecological Restoration, Department of Ecology and Evolutionary Biology, Biological Sciences West, University of Arizona, Tuscon, AZ 85721, USA*

M.F. M. FENNER *Biodiversity and Ecology Division, School of Biological Sciences, University of Southampton, Bassett Crescent East, Southampton SO16 7PX, UK*

B.W.F. **B.W. FERRY** *School of Biological Sciences, Royal Holloway, University of London, Egham Hill, Egham TW20 0EX, UK*

E.A.F. **E.A. FITZPATRICK** *Department of Plant and Soil Science, University of Aberdeen, Cruickshank Building, St Machar Drive, Aberdeen AB24 3UU, UK*

T.L.F. **T.L. FLEISCHNER** *Environmental Studies Program, Prescott College, 220 Grove Avenue, Prescott, AZ 86301, USA*

V.F. **V. FORBES** *Department of Life Sciences and Chemistry, Roskilde University, PO Box 260, DK-4000 Roskilde, Denmark*

N.R.F. **N.R. FRANKS** *Department of Biology and Biochemistry, University of Bath, Claverton Down, Bath BA2 7AY, UK*

R.P.F. **R.P. FRECKLETON** *Schools of Environmental and Biological Sciences, University of East Anglia, Norwich NR4 7TJ, UK*

W.G. **W. GABRIEL** *Zoologisches Institut der Ludwig-Maximilians-Universität, Postfach 20 21 36, D-80021 München, Germany*

J.H.C.G. **J.H.C. GASH** *Institute of Hydrology, Crowmarsh Gifford, Wallingford OX10 8BB, UK*

M.T.G. **M.T. GHISELIN** *California Academy of Sciences, Golden Gate Park, San Francisco, CA 94118, USA*

O.L.G. **O.L. GILBERT** *Department of Landscape, University of Sheffield, Sheffield S10 2TN, UK*

P.S.G. **P.S. GILLER** *Department of Zoology and Animal Ecology, University College Cork, Lee Maltings, Prospect Row, Ireland*

H.C.J.G. **H.C.J. GODFRAY** *Department of Biology, Imperial College of Science, Technology and Medicine, Silwood Park, Ascot SL5 7PY, UK*

F.B.G. **F.B. GOLDSMITH** *Department of Biology, University College London, University of London, Gower Street, London WC1 6BT, UK*

J.G. **J. GRACE** *Institute of Ecology and Resource Management, University of Edinburgh, Mayfield Road, Edinburgh EH9 3JU, UK*

M.C.G. **M.C. GRAHAM** *Department of Chemistry, University of Edinburgh, West Mains Road, Edinburgh EH9 3JJ, UK*

A.N.G. **A.N. GRAY** *Forest Science Department, Oregon State University, Corvallis, OR 97331, USA*

P.J.G. **P.J. GRUBB** *Department of Plant Sciences, University of Cambridge, Downing Street, Cambridge CB2 3EA, UK*

E.O.G. **E.O. GUERRANT JR** *Berry Botanic Garden, 11505 SW Summerville Avenue, Portland, OR 97219, USA*

S.J.H. **S.J. HALL** *School of Biological Sciences, The Flinders University of South Australia, GPO Box 2100, Adelaide 5001, Australia*

A.H. **A. HALLAM** *School of Earth Sciences, University of Birmingham, Edgbaston, Birmingham B15 2TT, UK*

P.H.H. **P.H. HARVEY** *Department of Zoology, University of Oxford, South Parks Road, Oxford OX1 3PS, UK*

M.H. **M. HASSALL** *School of Environmental Sciences, University of East Anglia, Norwich NR4 7TJ, UK*

J.B.H. **J.B. HEALE** *Division of Life Sciences, King's College London, University of London, Campden Hill Road, London W8 7AH, UK*

G.A.F.H. **G.A.F. HENDRY** *Biological Sciences, University of Dundee, Dundee DD1 4HN, UK*

M.O.H. **M.O. HILL** *ITE Monks Wood, Abbots Ripton, Huntingdon PE17 2LS, UK*

A.R.H. **A.R. HOELZEL** *Department of Biological Sciences, University of Durham, South Road, Durham DH1 3LE, UK*

J.J.H. **J.J. HOPKINS** *Joint Nature Conservation Committee, Monkstone House, Peterborough PE1 1JY, UK*

R.B.H. **R.B. HUEY** *Department of Zoology, University of Washington, Box 351800, Seattle, WA 98195, USA*

R.N.H. **R.N. HUGHES** *School of Biological Sciences, University of Wales, Bangor, Deiniol Road, Bangor LL57 2UW, UK*

L.D.H. **L.D. HURST** *Department of Biology and Biochemistry, University of Bath, Claverton Down, Bath BA2 7AY, UK*

H.A.P.I. **H.A.P. INGRAM** *Johnstonfield, Dunbog, Cupar, Fife KY14 6JG, UK*

M.I. **M. INGROUILLE** *Department of Biology, Birkbeck College, University of London, Malet Street, London WC1E 7HX, UK*

S.K.J. **S.K. JAIN** *Department of Agronomy and Range Science, University of California at Davis, Davis, CA 95616, USA*

G.D.J. **G. DE JONG** *Department of Plant Ecology and Evolutionary Biology, Universiteit of Utrecht, Padualaan 8, 3584 CH Utrecht, The Netherlands*

T.J.K. **T.J. KAWECKI** *Zoology Institute, University of Basel, Rheinsprung 9, CH-4051 Basel, Switzerland*

J.K. **J. KINDERLERER** *Department of Molecular Biology and Biotechnology, University of Sheffield, Sheffield S10 2TN, UK*

T.B.L.K. **T.B.L. KIRKWOOD** *Department of Geriatric Medicine, University of Manchester, Oxford Road, Manchester M13 9PT, UK*

J.C.K. **J.C. KOELLA** *Department of Zoology, University of Aarhus, Universitetsparken B135, DK-8000 Aarhus C, Denmark*

W.D.K. **W.D. KOENIG** *Hastings Natural History Reservation, University of California at Berkeley, 38601 East Carmel Valley Road, Carmel Valley, CA 93924, USA*

J.Kz. **J. KOZLOWSKI** *Institute of Environmental Biology, Jagiellonian University, Oleandry 2a, 30-063 Krakow, Poland*

W.E.K. **W.E. KUNIN** *Ecology and Evolution Research Group, School of Biology, University of Leeds, Leeds LS2 9JT, UK*

J.L. **J. LANCASTER** *Institute of Ecology and Resource Management, University of Edinburgh, Mayfield Road, Edinburgh EH9 3JU, UK*

D.C.L. **D.C. LEDGER** *Institute of Ecology and Resource Management, University of Edinburgh, Mayfield Road, Edinburgh EH9 3JU, UK*

C.M.L. **C.M. LIVELY** *Department of Biology, Indiana University, Bloomington, IN 47405, USA*

J.Ly. **J. LLOYD** *Max-Planck-Institut für Biogeochemie, Sophienstrasse 10, D-07743 Jena, Germany*

J.W.L. **J.W. LLOYD** *School of Earth Sciences, University of Birmingham, Edgbaston, Birmingham B15 2TT, UK*

S.P.L. **S.P. LONG** *Department of Biological Sciences, John Tabor Laboratories, University of Essex, Wivenhoe Park, Colchester CO4 3SQ, UK*

L.L.L. **L.L. LOOPE** *US Geological Survey, Biological Resources Division, Pacific Islands Ecosystem Research Center, Haleakala National Park Field Station, PO Box 369, Makawao, HI 96768, USA*

T.F.C.M. **T.F.C. MACKAY** *Department of Genetics, North Carolina State University, Box 7614, Raleigh, NC 27695, USA*

A.B.M. **A.B. MACKENZIE** *Scottish Universities Reactor Centre, East Kilbride, Glasgow G75 0QF, UK*

A.E.M. **A.E. MAGURRAN** *School of Environmental and Evolutionary Biology, Bute Building, University of St Andrews, St Andrews KY16 9TS, UK*

J.L.B.M. **J.L.B. MALLET** *Galton Laboratory, Department of Biology, University College London, University of London, 4 Stephenson Way, London NW1 2HE, UK*

A.M.M. **A.M. MANNION** *Department of Geography, University of Reading, Whiteknights, Reading RG6 6AB, UK*

N.M. **N. MAWDSLEY** *Ecology and Evolution Research Group, School of Biology, University of Leeds, Leeds, LS2 9JT, UK*

A.M. **A. McDONALD** *Geography Department, University of Leeds, Woodhouse Lane, Leeds LS2 9JT, UK*

J.F.R.M. **J.F.R. McILVEEN** *Environmental Science Department, Institute of Environmental and Natural Sciences, Lancaster University, Lancaster LA1 4YQ, UK*

L.R.M. **L.R. McMAHAN** *The Berry Botanic Garden, 11505 SW Summerville Avenue, Portland, OR 97219, USA*

G.F.M. **G.F. MEDLEY** *Department of Biological Sciences, University of Warwick, Coventry CV4 7AL, UK*

J.M. **J. MILBURN** [Deceased] *Department of Botany, University of New England, Armidale, New South Wales 2351, Australia*

M.Mi. **M. MILINSKI** *Zoology Institute, University of Bern, Wohlenstrasse 50a, CH-3032 Hinterkappelen, Switzerland*

M.M. **M. MOGIE** *Department of Biology and Biochemistry, University of Bath, Claverton Down, Bath BA2 7AY, UK*

A.P.M. **A.P. MØLLER** *Laboratoire d'Ecologie, CNRS URA 258, Université Pierre et Marie Curie, Bât. A, 7ème étage, 7 quai St Bernard, Case 237, F-75252 Paris Cedex 05, France*

J.B.M. **J.B. MONCRIEFF** *Institute of Ecology and Resource Management, University of Edinburgh, Mayfield Road, Edinburgh EH9 3JU, UK*

P.D.M. **P.D. MOORE** *Division of Life Sciences, King's College London, University of London, Campden Hill Road, London W8 7AH, UK*

J.I.L.M. **J.I.L. MORISON** *Department of Biological Sciences, University of Essex, Wivenhoe Park, Colchester CO4 3SQ, UK*

P.B.M. **P.B. MOYLE** *Department of Wildlife, Fish, and Conservation Biology, University of California at Davis, Davis, CA 95616, USA*

C.E.M. **C.E. MULLINS** *Department of Plant and Soil Science, University of Aberdeen, Cruickshank Building, Aberdeen AB24 3UU, UK*

C.J.N. **C.J. NAGELKERKE** *Institute for Systematics and Population Biology, University of Amsterdam, Kruislaan 320, 1098 SM Amsterdam, The Netherlands*

M.O'C. **M. O'CONNELL** *Palaeoenvironmental Research Unit, Department of Botany, National University of Ireland, Galway, Ireland*

P.O. **P. OLEJNICZAK** *Institute of Environmental Biology, Jagiellonian University, Oleandry 2a, 30-063 Krakow, Poland*

I.O. **I. OLIVIERI** *Institut des Sciences de l'Evolution, Université Montpellier II, Place Eugene Bataillon, 34095 Montpellier, Cedex 05, France*

J.R.P. **J.R. PACKHAM** *School of Applied Sciences, University of Wolverhampton, Wulfruna Street, Wolverhampton WV1 1SB, UK*

P.P. **P. PAMILO** *Department of Genetics, Uppsala University, Box 7003, S-750 07 Uppsala, Sweden*

H.E.H.P. **H.E.H. PATERSON** *Department of Entomology, University of Queensland, St Lucia, Queensland 4072, Australia*

A.P. **A. PENTECOST** *Division of Life Sciences, King's College London, University of London, Campden Hill Road, London W8 7AH, UK*

G.F.P. **G.F. PETERKEN** *Beechwood House, St Briavels Common, Lydney GL15 6SL, UK*

T.D.P. **T.D. PRICE** *Department of Biology, University of California at San Diego, La Jolla, CA 92093, USA*

D.L.J.Q. **D.L.J. QUICKE** *Department of Biology, Imperial College of Science, Technology and Medicine, Silwood Park, Ascot SL5 7PY, UK*

P.B.R. **P.B. RAINEY** *Department of Plant Sciences, University of Oxford, South Parks Road, Oxford OX1 3RB, UK*

D.J.R. **D.J. RANDALL** *Department of Zoology, University of British Columbia, Vancouver, British Columbia V6T 1Z4, Canada*

J.D.R. **J.D. REYNOLDS** *School of Biological Sciences, University of East Anglia, Norwich NR4 7TJ, UK*

D.N.R. **D.N. REZNICK** *Department of Biology, University of California, Riverside, CA 92521, USA*

H.R. **H. RICHNER** *Zoology Institute, University of Bern, Wohlenstrasse, CH-3032 Hinterkappelen, Switzerland*

R.H.R. **R.H. ROBICHAUX** *Department of Ecology and Evolutionary Biology, University of Arizona, Tucson, AZ 85721, USA*

D.A.R. **D.A. ROFF** *Department of Biology, McGill University, 1205 Dr Penfield Avenue, Montreal, Quebec H3A 1B1, Canada*

M.R.R. **M.R. ROSE** *Department of Ecology and Evolutionary Biology, University of California, Irvine, CA 92697, USA*

M.L.R. **M.L. ROSENZWEIG** *Department of Ecology and Evolutionary Biology, University of Arizona, Tucson, AZ 85721, USA*

G.R. G. RUSSELL *Institute of Ecology and Resource Management, University of Edinburgh, West Mains Road, Edinburgh EH9 3JG, UK*

M.W.S. M.W. SABELIS *Institute for Systematics and Population Biology, University of Amsterdam, Kruislaan 320, 1098 SM Amsterdam, The Netherlands*

P.S.H. P. SCHMID-HEMPEL *Experimental Ecology, ETH Zurich, ETH-Zentrum NW, CH-8092 Zurich, Switzerland*

J.G.S. J.G. SEVENSTER *Institute of Evolutionary and Ecological Sciences, University of Leiden, PO Box 9516, NL-2300 RA Leiden, The Netherlands*

B.S. B. SHORROCKS *Ecology and Evolution Research Group, School of Biology, University of Leeds, Leeds LS2 9JT, UK*

R.M.S. R.M. SIBLY *School of Animal and Microbial Sciences, University of Reading, Whiteknights, Reading RG6 6AJ, UK*

K.A.S. K.A. SMITH *Institute of Ecology and Resource Management, University of Edinburgh, West Mains Road, Edinburgh EH9 3JG, UK*

R.H.S. R.H. SMITH *Department of Biology, University of Leicester, University Road, Leicester LE1 7RH, UK*

T.A.S. T.A. SPIES *USDA Forest Service, Pacific Northwest Forest Research Station, 3200 West Jefferson Way, Corvallis, OR 97331, USA*

J.M.S. J.M. STARCK *Institute of Zoology and Evolutionary Biology, University of Jena, Erberstrasse 1, D-07743 Jena, Germany*

S.C.S. S.C. STEARNS *Zoology Institute, University of Basel, Rheinsprung 9, CH-4051 Basel, Switzerland*

J.M.Sz. J.M. SZYMURA *Department of Comparative Anatomy, Jagiellonian University, ul Ingardena 6, 30-060 Krakow, Poland*

J.H.T. J.H. TALLIS *School of Biological Sciences, University of Manchester, Oxford Road, Manchester M13 9PT, UK*

K.T. K. THOMPSON *NERC Unit of Comparative Plant Ecology, Department of Animal and Plant Sciences, University of Sheffield, Sheffield S10 2TN, UK*

J.R.G.T. J.R.G. TURNER *Ecology and Evolution Research Group, School of Biology, University of Leeds, Leeds LS2 9JT, UK*

M.V. M. VALERO *Laboratoire de Génétique et Evolution des Populations Végétale, Université de Lille 1, F-59655 Villeneuve d'Ascq Cedex, France*

J.V.A. J. VAN ALPHEN *Institute of Evolutionary and Ecological Sciences, University of Leiden, PO Box 9516, NL-2300 RA Leiden, The Netherlands*

P.R.V.G. P.R. VAN GARDINGEN *Institute of Ecology and Resource Management, University of Edinburgh, West Mains Road, Edinburgh EH9 3JG, UK*

L.M.V.V. L.M. VAN VALEN *Department of Ecology and Evolution, University of Chicago, 1101 East 57th Street, Chicago, IL 60637, USA*

Y.V. Y. VASARI *Department of Ecology and Systematics, University of Helsinki, PO Box 7 (Unioninkatu 44), Fin-00014 University of Helsinki, Finland*

P.I.W. P.I. WARD *Zoologisches Museum, University of Zurich-Irchel, Winterthurerstrasse 190, CH-8057 Zurich, Switzerland*

P.H.W. P.H. WARREN *Department of Animal and Plant Sciences, University of Sheffield, Sheffield S10 2TN, UK*

W.W.W. W.W. WEISSER *Zoology Institute, University of Basel, Rheinsprung 9, CH-4051 Basel, Switzerland*

T.C.W. T.C. WHITMORE *Department of Geography, University of Cambridge, Downing Place, Cambridge CB2 3EN, UK*

A.J.W. A.J. WILLIS *Department of Animal and Plant Sciences, University of Sheffield, Sheffield S10 2TN, UK*

S.N.W. S.N. WOOD *Mathematical Institute, North Haugh, St Andrews KY16 9SS, UK*

S.R.J.W. S.R.J. WOODELL *Wolfson College, Linton Road, Oxford OX2 6UD, UK*

Preface

Following the successful launch of *The Encyclopedia of Ecology & Environmental Management* in 1998, we have been persuaded to prepare a couple of shorter and more focused versions. This one, *Blackwell's Concise Encyclopedia of Ecology*, takes edited versions of entries from the parent volume that are concerned with all aspects of ecology. A sister version does the same for environmental management.

The aim has been to produce a concise version that meets the day-to-day needs of students, teachers and professionals working in this area. The entries culled from the full encyclopedia will, in general, be briefer than the originals, but there has still been an attempt, in many of the key terms, to go beyond the few-line definitions that are the hallmark of dictionaries to a more in-depth and critical appraisal, often with a historical commentary. Because of their importance in ecology, we also include a large number of definitions of statistical terms. This is again something that distinguishes what follows from what might be expected of a dictionary of ecological terms.

This concise version contains 1500 entries with extensive cross-referencing. Any relevant headword that is mentioned within the text of any other appears in SMALL CAPITALS. In addition, obviously related headwords are linked by *See also* at the end of entries. For brevity we have not included references, but there are a few figures and a series of helpful tables. Other conventions on hyphenation, alphabetical order, taxonomic nomenclature and abbreviations and acronyms are as in the parent work.

The initials of contributors are given at the end of each entry and their details are given on pp. vii–xii. The success of the *Encyclopedia* has been due to the joint efforts of these many contributors and this concise version continues to owe much to them. I would also like to take this opportunity to record thanks to my secretary, Samantha Giles, and Blackwell's staff for excellent support in bringing this work into being. The success of our efforts is, of course, for users to determine and as with the *Encyclopedia* we would appreciate any comments, even negative ones, that might help us improve future editions.

PETER CALOW
Sheffield, 1999

A

abiotic factors (physical factors) A descriptive collective term for components of the physical environment, for example TEMPERATURE, moisture and light. *Cf.* BIOTIC FACTORS, which usually refers to other living organisms, for example competitors, natural enemies and host plants. [J.S.B.]

absorption efficiency *See* ASSIMILATION EFFICIENCY.

abundance The availability of a resource, or numbers in a population, often described in qualitative terms such as 'rare' or 'common'. [R.H.S.]

acaricide A BIOCIDE intended to kill ticks and mites. *See also* INSECTICIDES; PESTICIDE. [P.C.]

acclimation The widespread phenomenon whereby living organisms adjust to the present environmental conditions, and in doing so, enhance their probability of survival. These adjustments are on time-scales of less than one generation, and may involve changes in physiological processes and structure. [J.G. & J.S.B.]

acclimatization *See* ACCLIMATION.

accuracy/precision These terms commonly refer to how well a particular measurement has been made and are often used interchangeably, although strictly they do have slightly different meanings. Accuracy is the closeness of a measured or computed value to its true value; precision is the closeness of repeated measurements of the same quantity. Statistically, accuracy is determined by the size of a SAMPLE while precision is determined by the way in which the sample is taken. [R.C.]

acid rain and mist Rain is naturally acidic, containing carbonic acid formed by the solution of atmospheric carbon dioxide (CO_2), and has a pH of about 5.6. More acidic rain (lower pH) may be formed naturally from biogenic sources of sulphur-containing gases, such as dimethyl sulphide ($(CH_3)_2S$), which is released by marine PHYTOPLANKTON. These gases are oxidized to sulphur dioxide (SO_2), and ultimately to sulphuric acid (H_2SO_4), which is incorporated into cloud and raindrops. In polluted regions, anthropogenic emissions of sulphur and nitrogen oxides from fossil fuel combustion are oxidized in the ATMOSPHERE to sulphuric and nitric acids, which cause widespread ACIDIFICATION of cloud and rainwater. [J.N.C.]

acidification Reduction in the pH of the environment, mainly due to human actions; for example, ACID RAIN AND MIST and acid mine drainage. [P.C.]

acidophilic Describing plants that are confined to, or more common on, acid, but not necessarily calcium-deficient, soil. [J.R.E.]

acidophobic Describing plants that are excluded from acid soil, usually by sensitivity to high concentrations of iron, manganese and aluminium, inadequacy of nitrate supply, low concentration of phosphorus and general mineral nutrient deficiency. [J.R.E.]

acquired character A change in an organism due to environmental influence. It may be adaptive (due to selection for adaptative PHENOTYPIC PLASTICITY) or maladaptive. Lamarckian inheritance presumed that acquired characters were heritable. [P.C.]

acrotelm The upper functional SOIL layer in a MIRE, in which plant litter is transformed into PEAT and beneath the base of which the WATER TABLE never sinks in an intact system. [H.A.P.I.]

active dispersal The process by which organisms actively move away from one another as a means of increasing FITNESS. [R.H.S.]

acute tests Testing, often referring to ECOTOXICOLOGY, for responses that are immediately debilitating; i.e. usually lethal over a short time-span (day(s)), and so involve relatively high concentrations of test substance. *Cf.* CHRONIC TESTS. [P.C.]

adaptation A term generally used to mean within-species evolutionary response to a particular, often new environment. The implication is that NATURAL SELECTION has favoured alleles that gave their carriers an advantage in the specified environment. [R.M.S.]

adaptive landscape

1 The mean FITNESS of a population as a function of gene frequencies.

2 The mean fitness of a population as a function of the mean PHENOTYPE.

While these are formal definitions, the concept is also used heuristically to describe positions of alternative equilibria with respect to the GENOTYPE or phenotype frequencies. Empirical descriptions of adaptive surfaces are often based on estimates of individual fitnesses rather than mean population fitness. [T.D.P.]

adaptive radiation The evolutionary process by which an ancestral species gives rise to an array of descendant species exhibiting great ecological, morphological or behavioural diversity. [R.H.R.]

additive experiments A design for experiments on interspecific plant COMPETITION, in which a constant density of individuals of species A is grown in MONOCULTURE and with various numbers of species B (e.g. yield is measured when a crop plant, sown at standard density, is grown with various densities of weeds). Thus the total number of individuals (species A and B together) differs from treatment to treatment, which means that any effects of INTERSPECIFIC COMPETITION are confounded with the effects of increased total POPULATION DENSITY. *See also* DE WIT REPLACEMENT SERIES; EXPERIMENTAL DESIGN. [M.J.C.]

additive genetic variance The additive GENETIC VARIANCE V_A is defined in two ways: in statistical empirical QUANTITATIVE GENETICS it is the variance of the BREEDING VALUES in the population, whereas in theoretical quantitative genetics it is that part of the genetic variance that is explained by linear regression of genotypic values on number of A_1 ALLELES in the diploid GENOTYPE. [G.D.J.]

adoption The providing of all PARENTAL CARE to an immature individual by an individual that is not one of its biological parents. [C.B.]

aerobic respiration A metabolic process, occurring in the presence of oxygen, in which organic substances are broken down to yield energy with molecular oxygen acting as the final electron acceptor. [V.F.]

aestivation A state of inactivity occurring in some animals, for example lungfish, during prolonged periods of DROUGHT. Physiological processes slow down when an organism aestivates. *Cf.* DORMANCY. [P.O.]

affinity index An index of similarity (A) of the species composition of two samples. $A = C/\sqrt{(a + b)}$, where a and b are the numbers of species unique to each SAMPLE and C is the number of species common to both. *See also* SAMPLING. [P.C.]

afforestation The ESTABLISHMENT of FOREST either by natural processes, i.e. during SUCCESSION, or by planting. *Cf.* DEFORESTATION. *See also* FORESTATION. [P.C.]

age and size at maturity From the point of view of NATURAL SELECTION for reproductive success, maturation is seen as dividing a life into two phases: prior preparation and subsequent fulfilment. Many organisms delay maturity, and variation in both age and size at maturity can be detected among species, among populations within species, and among individuals within populations. [S.C.S.]

age class Grouping of the organisms within a POPULATION by age instead of by life stage (e.g. in the construction of a LIFE TABLE). [A.J.D.]

age structure The relative numbers of individuals within different AGE CLASSES in a POPULATION. Age structure changes according to the birth and death rates acting at different ages currently and in the recent past of a population. Age structure is often summarized in a LIFE TABLE. *See also* STABLE AGE DISTRIBUTION. [R.H.S.]

ageing Ageing entails physiological deterioration with chronological age among the adults of most species of animals and plants, even when they are kept under good conditions. [M.R.R.]

aggregated distribution A term used to describe the spatial distribution of the individuals in a POPULATION. Other equivalent terms are contagious distribution, clumped distribution or OVERDISPERSION. When individuals in a population follow an aggregated distribution, the variance (s^2) of individuals per sample is greater than the MEAN number ($\bar{x}$) of individuals per sample ($s^2 > \bar{x}$). [B.S.]

aggregation The term aggregation has several meanings in ecology. The first is a simple description of a group of individuals that are close together in space and is synonymous with other general terms such as clump or cluster, or more specific ones such as shoal or COLONY.

The second meaning is a description of the DISPERSION of individuals in space. If a POPULATION is distributed randomly in space then the MEAN and variance of samples from that population will be equal. If the variance/mean ratio is greater than 1 then this implies aggregation of individuals (i.e. they are more contagious, clustered or clumped than random). [C.D.]

aggregation model of competition Many insects exploit resources which are patchy, consisting of small, separate units, and which are ephemeral in the sense that they persist for only one or two generations. Such resources can include fruit, fungi, sap flows, decaying leaves, flowers, dung, carrion, seeds, dead wood and small bodies of water held in parts of terrestrial plants (i.e. PHYTOTELMATA). This general view of insect ecology inspired the aggregation model of competition that allows a competitively inferior species to survive in PROBABILITY REFUGES. These are patches of resource (a single fungus, fruit, etc.) with no or a few superior competitors, that arise because the competing stages (usually larvae) have an AGGREGATED DISTRIBUTION across the patches. [B.S.]

air, chemical and physical properties Air is largely nitrogen (78% by volume) and oxygen (21%), with important but small concentrations of carbon dioxide (CO_2; 0.035%) and very low concentrations of several other gases (Table A1). The water vapour content is variable, with a fractional volume in the range 0.001–0.03. Common pollutant gases (*see* AIR POLLUTION) near ground level vary according to the presence of local sources and the meteorological conditions. Concentrations that become damaging to plants are: ozone 300 ppbv; nitrogen oxide ($NO + NO_2$) 30 μg m^{-3}; sulphur dioxide (SO_2) 30 μg m^{-3}.

Table A1 The composition of air, based on the US Standard Atmosphere 1976, specified here for sea-level dry air.

Gas species	Molecular weight	Fractional volume
N_2	28.0134	0.78084
O_2	31.9988	0.209476
Ar	39.948	0.00934
CO_2	44.01	0.00035
Ne	20.183	0.00001818
He	4.003	0.00000524
Kr	83.80	0.00000114
Xe	131.30	0.000000087
CH_4	16.04	0.000002
H_2	2.016	0.0000005

aerosols, including dust and pollens, are very variable. [J.G.]

air pollution The existence of CONTAMINANTS in the air at levels that interfere with human health and/or ecological systems. [P.C.]

air quality Description of the extent to which the ATMOSPHERE in a defined locale is contaminated and polluted. [P.C.]

alarm calls Visual, oral or olfactory signals emitted by one individual to warn others in the group of danger. [P.C.]

albedo The shortwave (300–1500 nm) reflectance of a natural surface, also known as the reflection coefficient, obtained by measuring the incident (S_i) and reflected (S_r) solar irradiance over the surface (land or water). [J.G.]

algal bloom Dense populations of free-floating algae, often imparting a distinctive colour and odour to the water body. BLOOMS occur in marine and fresh waters and result from a period of intense, often monospecific growth in response to favourable nutrition and light. [A.P.]

alien A non-indigenous organism. [M.J.C.]

Allee effect A special type of DENSITY DEPENDENCE which occurs in low-density populations. It was first identified by W.C. Allee in 1931. Conventionally, density dependence is negative; as DENSITY decreases, survival and reproduction increase due to the lessening of COMPETITION between individuals. However, there may be a point at which reducing density actually decreases survival or reproduction—the Allee effect. [C.D.]

allele frequency Genetic variation in a population at a locus A implies the presence of more than one allele at that locus. For two ALLELES A_1 and A_2, the frequency (p) of allele A_1 is given by:

$$p = [(2 \times \text{number of } A_1A_1 \text{ homozygotes}) + (\text{number of } A_1A_2 \text{ heterozygotes})] / (2 \times \text{total number of individuals})$$

for diploid organisms. [G.D.J.]

alleles Two or more forms of a gene occupying the same locus on a chromosome. [P.O.]

allelopathy A form of INTERFERENCE COMPETITION by means of chemical signals; i.e. compounds produced by one species of plant which reduce the germination, establishment, growth, survival or fecundity of other species. [M.J.C.]

Allen curve A method for estimating secondary production in which N_t, the number of the survivors (usually in a COHORT) at age t is plotted against W_t, the mean mass of individuals at age t. The area under the curve between W_{t1} and W_{t2} gives the production in the interval between $t1$ and $t2$. *See also* PRODUCTIVITY; SECONDARY PRODUCTIVITY. [S.J.H.]

Allen's rule Allen's rule is based on the observation that anatomical extremities in endothermic animals become progressively shorter as latitude increases. This pattern has been attributed to the proportional relationship between heat loss and body surface area, particularly heat loss from the periphery of the vascular system. Another similar but more contentious rule (BERGMANN'S RULE) also uses heat loss as an explanation for latitudinal increases in body mass of certain widely distributed mammals, such as deer. This has been attributed to the fact that larger animals experience a proportional decrease in heat loss as body mass increases because of the relationship between surface area and body mass. *See also* ALLOMETRY. [R.C.]

allochthonous Describing organic matter not generated within the COMMUNITY. [P.O.]

allogenic This term was first applied to SUCCESSIONS which take place as a result of the action of forces originating outside the system, for example the sinking of land relative to the sea, or deposition of silt in a lake, and a contrast was made with AUTOGENIC successions that arise as a result of the effects of plants on the HABITAT. Later, these terms were applied to factors rather than successions, and suggested that both allogenic and autogenic factors are present in all successions. Many successions are initiated by an allogenic event, for example landslide, deposition of a sandbank or retreat of a glacier, but are 'driven' by autogenic processes, for example nitrogen fixation or change in soil pH. [P.J.G.]

allometric growth The kind of GROWTH in which the proportions of various body parts change. During allometric growth one measurement (y) increases non-linearly with the increase of another measurement (x). [J.Kz.]

allometry Organisms that differ in size or mass also differ in the relative size of different organs, in the amount of time they spend performing particular behaviours, or when in the life cycle different events occur. Such size-related differences are termed 'allometric' and they can arise for a number of reasons. [P.H.H.]

allopatric speciation The differentiation of geographically isolated populations into distinct species. [V.F.]

allopolyploid An organism with more than two chromosome sets which are derived from different species (*cf.* AUTOPOLYPLOID). Allopolyploidy may be caused by multiplication of chromosome sets in a hybrid or by fusion of unreduced GAMETES of two different species. These processes are possible mechanisms of SPECIATION. [P.O.]

alluvial Pertaining to SEDIMENT transported and deposited by a flowing RIVER (i.e. alluvium). [P.C.]

alpha diversity *See* DIVERSITY, ALPHA, BETA AND GAMMA.

alpine zone The alpine vegetation zone lies between the upper limit of tree growth and the snow line. Within the alpine zone vegetation communities vary depending on altitude, aspect, exposure, snow cover and water availability. [R.M.M.C.]

alternation of generations The alternation of two different reproductive processes in the LIFE CYCLE of an organism. It is common and highly variable in its expression among animals, plants and fungi. [D.E.]

altitudinal zonation Vegetation shows a marked ZONATION with altitude because of changing TEMPERATURE and rainfall. In some ways the altitudinal zonation mirrors the latitudinal zonation (*see* LATITUDINAL GRADIENTS OF DIVERSITY) from the tropics to the polar regions, but the effects of both frost and the light regime are quite different between the tropics and other latitudes. [M.I.]

altricial Altriciality in chicks is characterized by a low functional maturity of tissues but high growth rates, closed eyes at birth and intense PARENTAL CARE for the young. The term altricial derives from the Latin *altrix*, wet-nurse. [J.M.S.]

altruism *See* KIN SELECTION.

ambient Of, or relating to, surrounding environmental conditions. [P.C.]

amensalism An interaction, usually between two species, in which one has a detrimental effect upon the other, while the other has no effect upon the first species. Amensalism is an extreme form of ASYMMETRIC COMPETITION. [B.S.]

amixis ASEXUAL REPRODUCTION in which progeny are produced mitotically, receiving a copy of the maternal genome. Vegetative proliferation is sometimes described as amictic although there is disagreement over whether it is a process of reproduction or of growth. [M.M.]

anadromous Describing a specific annual migratory pattern in fish when they move from marine to freshwater environments. [R.C.]

anaerobic metabolism Metabolism in the absence of oxygen. [V.F.]

anagenesis Evolutionary change within a LINEAGE. It is often contrasted with CLADOGENESIS, evolutionary change that results in SPECIATION and the branching of phylogenetic trees. [D.A.B.]

analogy The term analogy is used to classify characters that have evolved to perform a similar function in distantly related taxa. Birds' wings and insects' wings are a classical example of analogy because the two types of wings serve the same function (flying), are superficially similar morphologically (elongated, elliptical, thin, movable, appendage-like extensions), and they are found in distantly related taxa. [D.R.B.]

analysis of covariance (ANCOVA) A procedure closely related to that of ANALYSIS OF VARIANCE in which a group of independent variables (y) are tested for homogeneity of their means, after they have been adjusted for the group's differences in the independent variable (x). The analysis therefore involves the comparison of several regression lines. Slopes are assumed equal and y-intercepts are compared. [B.S.]

analysis of variance (ANOVA) A technique for comparing the difference between several SAMPLE means by analysing the variance in the total data. It is a parametric technique (*see* PARAMETRIC STATISTICS) and assumes that observations are normally distributed (*see* NORMAL DISTRIBUTION), with approximately equal variances in all samples. If these two conditions are not met they can frequently be obtained by TRANSFORMATION OF DATA. [B.S.]

ancient woodland/forest In Britain, WOODLAND which existed before 1600 or before 1700 depending on authors. Some ancient woodland has since been cleared and some survives in the modern landscape, having existed continuously since 1600. All PRIMARY WOODLAND is ancient woodland, but ancient woodland also includes SECONDARY WOODLAND originating before 1600. [G.F.P.]

androdioecy A species is said to be androdioecious when some or all populations of the species contain both male (hence the prefix 'andro') and HERMAPHRODITE individuals (broad sense) (hence the suffix 'dioecious', referring to the existence of two sexes). *Cf.* GYNODIOECY. [I.O.]

aneuploidy The usual definition is possession of a chromosome number that is not an exact multiple of the usual haploid (HAPLOIDY) number. However, some individuals may have a reduced number of chromosomes through fusions, without having a reduced number of arms. Such individuals are not described as aneuploids. Thus a better definition might be possession of a number of chromosomal arms that is not an exact multiple of the usual haploid number of arms ('fundamental number'). [I.O.]

angiosperms Flowering plants (angiosperms) are major components of the world's flora and its VEGETATION. Considered the culmination of evolution in the plant kingdom, they originated about 130 million years ago and evolved very rapidly, leading to an enormous progressive increase in plant diversity to the present day. The angiosperms are seed-producing vascular plants of extremely varied form; many are valuable as sources of food and other commodities. Dominant in virtually all plant communities—the exceptions being sphagnum BOGS and the

BOREAL region—they characterize biomes and, although mostly terrestrial mesophytes, range from desert xerophytes to aquatics (both freshwater and seawater). Major vegetation types include: many kinds of FOREST (e.g. TROPICAL RAIN FOREST, DECIDUOUS FOREST, sclerophyll forest); SAVANNAH and parkland GRASSLANDS, STEPPES, PRAIRIES and llanos; WETLANDS and coastal vegetation including MANGROVES. [A.J.W.]

anisogamy The production of GAMETES of two different sizes. By definition, the smaller gametes are male, while the larger gametes are female. [I.O.]

anisometric Relating to, or characterized by, dissimilar proportions. Living things of different sizes are usually anisometric, even when organized on a similar pattern, because of mechanical or physiological CONSTRAINTS. For instance, if animals were isometric, the strength of their bones (roughly proportional to bone cross-sectional area) would only increase with the square of their body length; but because the body mass that must be supported by the limbs increases approximately with the cube of body length, the limbs of large land animals are, in fact, disproportionately massive. *See also* ALLOMETRY. [J.Kz.]

annual LIFE CYCLE over 1 year, characteristic of many plant species. Annuals are often semelparous (*see* SEMELPARITY), i.e. breed once at the end of the year and die. However, semelparous species may be subannual. [I.O.]

anoxia (hypoxia) Environmental conditions with no (or low) oxygen. [P.C.]

antagonistic pleiotropy hypothesis Different ALLELES may have effects on multiple characters, including opposed effects on components of Darwinian FITNESS. [M.R.R.]

Antarctic ozone hole The seasonal depletion of ozone in a large area of the ATMOSPHERE over Antarctica. *See also* OZONE HOLE; STRATOSPHERIC CHEMISTRY. [P.C.]

anthropic horizons *See* SOIL CLASSIFICATION.

anthropocentric Either: (i) describing animals or plants that have become strongly associated with humans; or (ii) reasons or explanations that put humans in the centre of things. [O.L.G.]

anthropomorphism Attributing human qualities, especially to animal behaviour and/or to the processes of nature. [P.C.]

aphotic zone The region of the OCEAN and deep LAKES in which there is no light penetration and, hence, no PHOTOSYNTHESIS. [J.L.]

apomixis The production of females from unfertilized EGGS, without meiosis. The offspring thus produced are genetically identical to their mother. [I.O.]

aposematic (warning) coloration Distasteful, toxic or otherwise potentially dangerous prey items are frequently brightly coloured and contrast with their background. [P.H.H.]

apostatic selection A type of FREQUENCY-DEPENDENT SELECTION in which the common phenotypes are selected against and the rare phenotypes are favoured, for example a predator prefers the most frequent prey, which then suffers higher mortality. Apostatic selection may lead to a stable distribution of frequences of phenotypes. [P.O.]

apparent competition An ecological situation where two species appear to show the reciprocal negative effects associated with interspecific competition (*see* SPECIES INTERACTIONS), but this is, in fact, the result of predation by a third species. [B.S.]

appetite Appetite is commonly defined in terms of desire for, or inclination towards, certain commodities or other stimuli, most often in the context of food or sex. [C.J.B.]

aquaculture Culturing aquatic organisms for commercial purposes, either in artificial systems—for example tanks or channels—or in nature. [P.C.]

arms races A continuing evolutionary interaction between species in which changes in one species exert a SELECTION PRESSURE for changes in the other interacting species. Arms races refer to changes occurring on an evolutionary time-scale rather than on the scale of individual lifetimes. These changes, analogous to a sequence in the development of weapons and defences by opposing military forces, characterize predator–prey, parasite–host and COMPETITION interspecific interactions in which defences evolved, for example, by the prey or host select for the evolution of new offensive strategies by the predator or parasite (*see* PREDATOR–PREY INTERACTIONS). Arms races driven by SEXUAL SELECTION have also been used to describe evolutionary changes in males and females of the same species. [V.F.]

arrhenotoky A haplo-diploid genetical system (*see* HAPLO-DIPLOIDY) in which males arise from unfertilized EGGS, females from fertilized eggs. As a consequence males only transmit the genome of the mother. [M.W.S. & C.J.N.]

artificial selection The deliberate choice of a select group of individuals to be used for breeding. The response to artificial selection is usually quite large, especially during the first generations of selection. This suggests that the response is due to GENETIC VARIANCE already present, rather than to new MUTATIONS. [I.O.]

artificial substrate A material placed in the environment by an experimenter so that patterns of colonization and COMMUNITY development can be observed. [S.J.H.]

asexual reproduction In a broad sense, any form of reproduction not causing rearrangement of maternal genes, so producing offspring that are genetically identical by descent. In this sense, asexual reproduction is synonymous with cloning (*see* CLONE) either by somatic division (agametic cloning) or by PARTHENOGENESIS (gametic cloning).

In a strict sense, however, 'asexual reproduction' is reproduction not involving sexually derived processes. Parthenogenesis is excluded because it is derived from sexual reproduction. In this sense, asexual reproduction is synonymous only with agametic cloning (*see also* VEGETATIVE REPRODUCTION). It involves successive phases of GROWTH and division of the body, the mechanism of division varying from fragmentation by exogenous environmental forces to endogenous FISSION or budding. The rare phenomenon of polyembryony is a form of fission or budding confined to the embryonic stage in the LIFE CYCLE. [R.N.H. & A.J.W.]

assemblage A general term for a collection of plants and/or animals, or the fossilized remains of these organisms in a geological SEDIMENT. Unlike the term 'COMMUNITY', it does not imply interrelationships between the organisms and, in the case of FOSSIL ASSEMBLAGES, many and various factors may have been involved in the assembling of the individuals. [P.D.M.]

assemblage zone A BIOSTRATIGRAPHIC UNIT that is defined on the basis of the total fossil content, all or part of which constitutes 'a natural assemblage or association which distinguishes it in biostratigraphic character from adjacent strata'. [M.O'C.]

assembly rules Compared with all the possible combinations of species, from the species pool of a region only certain 'permissible' combinations of species actually occur. These permissible combinations resist invasion by species that would transform them into 'forbidden' combinations. The forbidden combinations do not exist in nature because they would transgress one or more of three types of rule. These assembly rules are called: (i) compatibility rules; (ii) incidence rules; (iii) combination rules. [B.S.]

assimilation The absorption and use of simple NUTRIENTS and/or products of digestion in metabolism and building up the constituents of an organism. Assimilated energy is also called metabolizable energy (ME), i.e. that proportion of the digested energy that is not excreted as urinary waste but remains available to fuel resting metabolism, the animal's activities and the production of new somatic and reproductive tissues. It is measured as the difference between the amount or energy content of food consumed minus the sum of materials or energy defecated and excreted, or by adding the amount of energy lost during RESPIRATION to the amount incorporated due to production of new tissues. *See also* ENERGY BUDGET. [M.H.]

assimilation efficiency Assimilation EFFICIENCY is the amount of food assimilated expressed as a percentage of the amount of food consumed. However, assimilation is usually intended to include absorbed energy only (*cf.* ASSIMILATION), so this efficiency is more properly an absorption efficiency. *See also* ECOLOGICAL EFFICIENCY; ENERGY BUDGET. [M.H.]

association Similar to CORRELATION, but a more general term. For example, it applies to the relationship between qualitative information in a CONTINGENCY TABLE. [B.S.]

association hypothesis The argument that females are more likely to care for their young than males because they are more likely to be in a position to help the young at birth. [D.A.B.]

assortative mating A reproductive system in which matings are not at random, with respect to PHENOTYPE or GENOTYPE characteristics of mating pairs. [I.O.]

asymmetric competition COMPETITION, usually between a pair of species, in which the adverse effect of one species on the other is much greater than the reciprocal effect. The extreme situation is called AMENSALISM. [B.S.]

atmosphere The mainly gaseous envelope overlying the solid and liquid surface of a planet. [J.F.R.M.]

atmospheric pollution The addition of substances to the ATMOSPHERE that have a capacity to cause HARM to human health and/or ecological systems. It can involve the emission of chemicals from stacks, the emission of smoke and soot from chimneys, the emission of smoke, soot and chemicals from fire, explosions, leaks, etc. *See also* AIR POLLUTION. [P.C.]

attack rate *See* SEARCHING EFFICIENCY.

attenuation, light Diminution of light intensity with depth in the OCEAN and freshwater bodies due to absorption and scattering in the water column. [V.F.]

aufwuchs *See* BENTHIC HABITAT CLASSIFICATION.

autecology The ECOLOGY of individual species as opposed to the ecology of whole communities. Autecology includes behavioural, physiological and population ecologies. [M.H.]

autochthonous Describing material originating in its present position; for example BIOMASS (leading to SEDIMENT) that is formed from PHOTOSYNTHESIS in a freshwater lake or river. *Cf.* ALLOCHTHONOUS. [P.C.]

autogenic A term first applied to SUCCESSIONS that take place as a result of the effects of the plants on the HABITAT, and a contrast was made with allogenic successions, which result from the action of forces originating outside the system. Later applied to factors rather than successions. [P.J.G.]

automixis The production of females from unfertilized EGGS, with a normal meiosis producing four haploid pronuclei. The diploid number is then restored by the fusion either of two of the haploid products of meiosis, or of two genetically identical nuclei produced by the mitotic division of the haploid cell nucleus. Offspring differ from their mother, in contrast to APOMIXIS and endomitosis. The offspring are usually homozygous at some or all loci at which the parent was heterozygous. *See also* ASEXUAL REPRODUCTION; PARTHENOGENESIS. [I.O.]

autopolyploid An organism with more than two chromosome sets, all from the same ancestral species. Autopolyploidy can induce SYMPATRIC SPECIATION through postzygotic REPRODUCTIVE ISOLATION (e.g. a cross between an autotetraploid and its diploid ancestor will produce sterile triploids). Autopolyploids often have low fertility and may be rare in nature. [M.M.]

autotroph Simply, an organism that can make its own food: the term is derived from the Greek *auto*, self, and *trophos*, feeder. Autotrophs can exploit simple inorganic compounds, using carbon dioxide (CO_2) or carbonates as the carbon source for building their organic constituents. Two categories of autotroph are defined by the energy source used in the synthesis of their organic requirements from inorganic carbon species.

1 Photoautotrophs: these use light energy, and comprise all photosynthetic organisms, including most terrestrial plants, algae and photosynthetic bacteria (*see also* PHOTOSYNTHESIS).

2 Chemoautotrophs: these comprise bacteria that obtain energy from the oxidation of simple inorganic or 1-C organic compounds and can use the energy released to assimilate CO_2 and transfer the energy into organic compounds. For example, *Thiobacillus* species can obtain their energy by oxidizing hydrogen sulphide or elemental sulphur to sulphuric acid. *Cf.* HETEROTROPH. [S.P.L.]

a_x In a LIFE TABLE, the number of individuals in AGE CLASS x or at stage x (where x defines the order of the classes or stages). [A.J.D.]

B

BACI (before–after/control–impact) A SAMPLING design used in MONITORING, involving replicate samples before and after DISTURBANCE in each of an undisturbed control and putatively impacted location. An impact due to the disturbance (e.g. effluent to a river from a POINT SOURCE) will cause the difference in mean ABUNDANCE from before to after, in the impacted location, to differ from any natural change in the control.

However, there might still be natural temporal variations in the control and putatively impacted site that confound the interpretation. For example, following the release of an effluent from a pipe to a river, adult insects may arrive and lay eggs in the upstream, control location but not in the downstream location, purely by chance. As a result, BACI would suggest an apparent impact. Hence, the interpretation is complicated and any effect of the disturbance cannot be distinguished from the natural variation. Therefore, if there is an environmental impact it is *necessary* to demonstrate differences between control and putatively impacted sites. But, in itself, this is not *sufficient* evidence of impact (*see also* PSEUDOREPLICATION).

A possible solution to these problems is to replicate the sampling, for example between control and impacted sites in separate RIVERS, but this is rarely possible or even desirable for impacted sites. So only the control sites might be replicated, and this leads to asymmetrical sampling designs for which some statistical procedures are available. *See also* BEFORE–AFTER STUDIES. [P.C.]

background level The concentration of a substance in the environment that is not attributable to a particular human activity. The term often refers to the concentration produced by NATURAL phenomena; i.e. of natural substances such as metals that can be added to by human activities. [P.C.]

bactericide A BIOCIDE used to control or destroy bacteria. [P.C.]

Bailey's triple-catch method A method of estimating POPULATION SIZE from mark–recapture data, using the shortest sequence possible for an open population. *See also* CAPTURE–RECAPTURE TECHNIQUES. [L.M.C.]

balance of nature The idea that there are tendencies in the natural world towards equilibria in populations, communities, ecosystems, etc. An extreme form of this is the notion that there is some kind of natural design or predetermined goal states—a notion rejected by most ecologists. A less extreme concept is that there are dynamic steady states brought about by BIOLOGICAL CONTROL systems of various kinds. *See also* GAIA; TELEOLOGY. [P.C.]

balance of selective forces When two or more selective forces act on a population or a group of populations, a balanced GENETIC POLYMORPHISM may be maintained (*see* BALANCED POLYMORPHISM). Most theories rely on the concept of protection of an allele: a given allele at a locus is said to be protected if its frequency increases when rare (*see* ALLELE FREQUENCY; ALLELES). This concept is very close to that of the EVOLUTIONARILY STABLE STRATEGY (ESS) in evolutionary GAME THEORY. A polymorphism at a locus is protected if at least two alleles are protected. Examples of protected polymorphisms at a bi-allelic locus can arise through various mechanisms, depending on the situation. [I.O.]

balanced diet Some evidence suggests that animals may sometimes achieve balanced diets by what looks like 'nutritional wisdom', i.e. decision rules that result in an animal acquiring its range of necessary dietary components. [C.J.B.]

balanced polymorphism POLYMORPHISM is a form of genetic variation in which different morphs, suites of behaviour or life histories within a population or species are observed at frequencies too common to be maintained by MUTATION. [M.C.M.B.]

basal species A term used to denote those species in a FOOD WEB which are preyed upon, but which do not themselves have prey. Strictly speaking, all basal species should be AUTOTROPHS, but the incompleteness of food web data sets has sometimes led authors to mistakenly classify non-autotrophs as basal species. *See also* FEEDING TYPES, CLASSIFICATION. [S.J.H.]

baseline data Data describing some original, or 'normal', state of a system—for example POPULATION DENSITY, population AGE STRUCTURE, species composition, energy flow—that can be used as standard/control against which changes are judged. This can often suffer from the problem of PSEUDOREPLICATION. [P.C.]

basic reproductive rate (R_0) The expected lifetime production of EGGS by a newborn female. Also called net reproductive rate or net FECUNDITY rate, it can be calculated from COHORT data as:

$$R_0 = \Sigma L_x M_x$$

where L_x is age-specific survivorship and M_x is age-specific fecundity. It is frequently used to describe the production of female offspring by females, and as a FITNESS measure suitable for modelling EVOLUTION in a stable population where R_0 for female offspring must equal 1. *Cf.* INTRINSIC RATE OF INCREASE. *See also* DEMOGRAPHY; LIFE-HISTORY EVOLUTION; LIFE TABLE. [D.A.B.]

Batesian mimicry A form of MIMICRY in which one (or more) palatable species of animal exhibit the same colour pattern as an unpalatable species. [P.M.B.]

bathyal zone The bottom of the OCEAN between 200 m and about 4000 m depth, corresponding to the depth of the CONTINENTAL SLOPE and continental rise. The bathyal zone occupies approximately 16% of the submerged ocean floor. [V.F.]

bathypelagic The oceanic water column between about 1000 and 4000 m depth, seaward of the shelf-slope break. [V.F.]

beat sampling *See* SAMPLING METHODOLOGY/DEVICES.

bedload The particles being moved along the bed of a RIVER or stream. [P.C.]

before–after studies A method of impact assessment that uses the state of the ENVIRONMENT and ecological systems within it as a standard against which to judge the condition following development and/or release of EFFLUENTS/emissions. However, because ECOSYSTEMS are dynamic it cannot always be presumed that the properties of the system(s) before would have been like those after, even if there had been no impact. *See also* BACI. [P.C.]

behavioural ecology Behavioural ecology can be defined as the study of the EVOLUTION of behaviour in relation to ECOLOGY. Its principal tenet is that behaviour patterns and decision-making have evolved mainly by NATURAL SELECTION and are thus adaptive in relation to problems of survival and reproduction imposed by the organism's ENVIRONMENT. Behavioural ecology emerged as an identifiable discipline in Europe and the USA in the early 1970s. [C.J.B.]

benthic ecology *See* BENTHIC HABITAT CLASSIFICATION; ECOLOGY.

benthic habitat classification Organisms associated with the substrate–water interface of LAKES, RIVERS, ESTUARIES and SEAS form the BENTHOS. The benthos can be either phytobenthos (plant) or zoobenthos (animal), both of which are subdivided into macro- (visible with the naked eye) or micro- (visible with a microscope).

Historically, botanists developed a CLASSIFICATION of the phytobenthos, which is equally applicable to the zoobenthos. The rhizobenthos is anchored (or rooted) into or on the substrate, the haptobenthos is adnate to the surface and is synonymous with the more commonly used terms PERIPHYTON or *aufwuchs*. Haptobenthos can also be classified with respect to the substrate on which it grows, being EPILITHIC on rocks, epiphytic on plants, epipelic on sediments, epizoic on animals and episammic on sand.

In addition to the true benthos, organisms found deeper in the sediments are hyporheic. The hyporheos consists of the interstices between the particles composing the substrate. [R.W.D.]

benthic–pelagic coupling The functional linkage between benthic and PELAGIC subsystems. The supply of organic matter from the PLANKTON to the BENTHOS is balanced by mineralization of ORGANIC DETRITUS by the benthic microbial community and release of dissolved metabolites to the overlying water. Interactions between the two subsystems are largely determined by the amount of TURBULENCE in the water column and can have important controlling influences on both pelagic and benthic community structure. [S.J.H. & V.F.]

benthos Benthos refers to all the attached, creeping or burrowing organisms that inhabit the bottom of RIVERS, LAKES and the SEA. The term is derived from Greek, meaning 'depth of the sea'. [V.F.]

Bergmann's rule A rule proposed by the German zoologist, C. Bergmann, in the mid-19th century, that HOMEOTHERMS within a single closely related evolutionary line increase in SIZE along a GRADIENT from warm to cooler CLIMATES. This is usually explained in terms of surface area to volume ratios, which reduce as body size increases in geometrically similar organisms. Less surface to volume means less heat loss—an advantage in cold conditions but not in warm conditions. *See also* ALLOMETRY. [P.C.]

bet-hedging A reproductive STRATEGY in temporally varying environments. To demonstrate bet-hedging, suppose a population of annual organisms inhabits an environment where wet and dry years occur randomly, independently and with equal probability (0.5). An individual of genotype 1 produces four surviving offspring in a wet year, but can expect only 0.5 offspring in a dry year. An individual of genotype 2 produces three surviving offspring in a wet year, but leaves only one offspring in a dry year. Genotype 1 has the greater arithmetic average fitness (2.25 > 2), but genotype 2 has the greater geometric mean fitness ($3^{1/2} > 2^{1/2}$). Genotype 2 has a sufficiently smaller variance in fitness between wet and dry years that its geometric mean fitness, hence its expected long-term growth, exceeds that of genotype 1. The lower variance more than compensates for genotype 2's lower arithmetic mean fitness. Genotype 2's less extreme (i.e. bet-hedged) response to temporal environmental variation indicates that its frequency in the population should increase relative to that of genotype 1. [T.C.]

biennial plants PERENNIAL semelparous species (*see* SEMELPARITY) that may flower the first year or wait for many years before flowering, and then die. [I.O. & G.A.F.H.]

bimodal distribution A DISTRIBUTION having two MODES. [B.S.]

binary fission The division of a body into two approximately equal parts. [M.M.]

binomial classification The practice, due to Linnaeus (*see* CHARACTERS IN ECOLOGY), of describing taxa under a generic and a specific name. Thus, *Musca domestica* is the name of a fly, species *domestica*, belonging to the genus *Musca*. By convention, both names are italicized and the generic, but not the specific, name has an initial capital. [L.M.C.]

binomial (positive binomial) distribution A discrete PROBABILITY DISTRIBUTION that describes situations where observations can fall into one of two categories (e.g. heads or tails when spinning a coin, and male or female offspring in a clutch or family). It is sometimes referred to as the positive binomial distribution (*cf.* NEGATIVE BINOMIAL DISTRIBUTION). [B.S.]

binomial (sign) test A non-parametric test that may be used in lieu of the parametric paired *t* TEST to test for differences between MATCHED OBSERVATIONS, or to test for GOODNESS-OF-FIT. [R.P.F.]

bioaccumulation The progressive increase in a substance—usually in an organism or part of an organism—because the rate of intake via the body surface (*see* BIOMAGNIFICATION) or in food is greater than the output from active or passive removal processes. [P.C.]

bioassays Methods of analysis that use living tissues or whole organisms or collections of organisms to make quantitative and/or qualitative measurements of the amounts or activity of substances. [P.C.]

bioavailability The extent to which a XENOBIOTIC substance can be taken up into an organism from that in food or the surrounding environment. [P.C.]

biochemical oxygen demand (BOD) A measure of oxygen absorption from a sealed sample of water over a fixed period, usually five days (BOD_5). It is therefore an index of ORGANIC LOADING and microbial activity and hence water quality. [P.C.]

biocide A generic term for any substance that kills (or inhibits) organisms. [P.C.]

biocoenosis A term used, most frequently in the East European literature, to denote biotic communities of populations living together in a physically defined space at the same time. *See also* COMMUNITY; ZOOCOENOSIS. [M.H.]

biodegradation The processes of natural DECOMPOSITION/decay whereby synthetic and natural compounds are broken down to produce substances that can be used in biological renewal cycles. [P.C.]

biodiversity The number and variety of taxa in ecological systems ranging from parts of communities to ECOSYSTEMS, regions and the BIOSPHERE. There is a deep concern that human activities are leading to species' losses, i.e. reduction in biodiversity. [P.C.]

biodiversity gradients Life is more 'abundant' at the tropics than at the poles. This applies not only to the BIOMASS and number of individuals, but in many taxonomic groups to the number of species as well. The cause of this planetary 'biodiversity gradient' in SPECIES RICHNESS remains in doubt. [J.R.G.T.]

bioenergetics The study of energy flow through biological systems. It includes both the physiological studies of energy transfer during metabolic processes and ecological studies of the rates and efficiencies of energy transfer through organisms in different trophic levels. *See also* ECOLOGICAL ENERGETICS. [M.H.]

biofilm, in aquatic systems The organic layer that coats all underwater surfaces. It is a heterogeneous collection of largely heterotrophic bacteria, microfungi, protozoans and micrometazoans such as rotifers. It can also contain photosynthetic algae. There is usually a polysaccharide matrix. [P.C.]

biofouling Unwanted accumulation of BIOTA on water-covered surfaces or appliances (e.g. hulls of boats, pipes). [P.C.]

biogenic sorting A regular change in SEDIMENT median grain size caused by the activities of benthic organisms. [V.F.]

biogeochemical cycle Any of various natural cycles of elements, involving biological and geological compartments. *See also* CARBON CYCLE; NITROGEN CYCLE; PHOSPHORUS CYCLE; SULPHUR CYCLE. [P.C.]

biogeographical units The world may be divided up into areas of similar character in terms of the animals and plants (the BIOTA) present in them. These biogeographical units are based on similarity of composition in terms of the systematics (and hence evolutionary history) of the biota. The extent and boundaries of these units have been determined by changes in CLIMATE and the movement of continents, and the accompanying changes in the physical and climatic barriers to migration.

A distinction must be made between such biogeographical units and the concept of biomes, which merely reflect the ecological character (and particularly the spectrum of life forms) in any given situation. Thus the RAIN FOREST biome is represented in humid tropical West Africa and in the Amazon Basin, but the actual composition of evergreen tropical forest in those two areas is very different at the level of species and genera. They are accordingly placed in two different biogeographical units (the Palaeotropical kingdom and the NEOTROPICAL kingdom respectively).

Zoologists, in delimiting zoogeographical units, have tended to put emphasis on the mammals and birds as the larger and more obvious features of the fauna, in recognizing faunal subdivisions of the world's biota. Plant biologists, in drawing up comparable phytogeographical units, base them largely on the distribution of flowering plants (angiosperms). [W.G.C.]

biogeography Biogeography is concerned with the patterns of DISTRIBUTION of species over the face of the globe and understanding the origins and mechanisms which determine the distribution. Biogeography is also concerned with exploring the underlying changes (territorial expansions and contractions) that have taken place in the past or, increasingly, which may occur under CLIMATES of the future. [A.H. & G.A.F.H.]

bioindicator An organism or its metabolic system that is used to signal the presence of a CONTAMINANT and its effects. A bioindicator may be a microbe or metazoan; bacterium, fungus, plant or animal; a single-species or multispecies system. *See also* BIOMARKERS. [P.C.]

biological control The limitation of the abundance of living organisms by other living organisms. Biological control can refer to the naturally occurring REGULATION of plant or animal populations by HERBIVORES, predators or diseases (natural biological control), or (more commonly) to the manipulation of these natural enemies by humans. Applied biological control is used as a means of reducing populations of PEST species, either on its own, or in combination with other control methods as part of an INTEGRATED PEST MANAGEMENT (INTEGRATED CONTROL) system. [S.G.C.]

biological invasions Expansion in the distribution of certain species of plants, animals and microorganisms which are transported by humans and often competitively favoured by the DISTURBANCE around human settlements. [L.L.L.]

biological monitoring Monitoring of biological organisms, species or ECOSYSTEMS is carried out with the aim of assessing the degree to which observations meet our expectations. This is usually conducted in the context of concern about loss of species, quality of HABITAT, increases in the levels of pollutants or some other problem. However, it is also important to monitor the effects of management carried out on reserves and other protected areas and to ensure that what we do is what we intended and is cost-effective.

There are several words in current usage with different shades of meaning in the context of MONITORING. These do not always have precise definitions and their meanings overlap to a certain extent. Biological recording is usually without motive, important and widely practised, whereas monitoring, strictly speaking, is conducted with clear objectives, a standardized procedure and clear rules for stopping (called termination by some people). Surveys are a one-off recording exercise whilst, strictly speaking, surveillance refers to repeated surveys. CENSUS involves recording all individuals and usually information about RECRUITMENT (births) and MORTALITY (deaths). It not only applies to animals but also to plants with discrete individuals such as orchids and trees. [F.B.G.]

biological species concept Groups of actually or potentially interbreeding natural populations which are reproductively isolated from other such groups. [L.M.C.]

bioluminescence Light emission, often as flashes and without sensible heat production, produced by bacteria, fungi, plants and animals at a wide range of frequencies from ultraviolet to the red end of the spectrum. [V.F.]

biomagnification The process whereby the concentration of a XENOBIOTIC chemical in the tissues of organisms increases as the chemical moves up the FOOD CHAIN. It is the result of BIOACCUMULATION and bioconcentration. Hence, only small amounts of a toxic chemical may accumulate per head of plant-eating animals; but many of these are eaten by predators, which therefore accumulate more of the chemical per head. [P.C.]

biomarkers Effects in biological systems used to indicate the presence of contamination. [P.C.]

biomass The total mass of living material within a specified area at a given time. It may be difficult to separate living tissues from those that are strictly dead (such as the wood tissues of a tree), so the term normally includes such non-living materials. An alternative expression, especially used of plant biomass, is 'STANDING CROP', which again may include the standing dead material often found in vegetation such as GRASSLANDS. Ideally, a value for the biomass of an ecosystem should include root material and soil organisms, but difficulties in the estimation of these has resulted in the use of 'above ground' biomass or standing crop data being extensively used. Biomass can be split into its various components, such as plant biomass, or species biomass, or into layers such as canopy, trunk and root biomass. It may also be split into various trophic levels. [P.D.M.]

biome *See* BIOGEOGRAPHICAL UNITS.

biometeorology The study of the interactions between weather and life. [J.G.]

biometrics The application of mathematical (usually statistical) techniques to biological systems. [P.C.]

biomonitor Any system incorporating living organisms (ranging from microbes to vertebrates) that can be used to assess (often continuously) quality of the ENVIRONMENT. [P.C.]

biosphere All the organisms on the planet, and their environment, viewed as a system of interacting components. The biosphere may be regarded as a thin film on the planet's surface, consisting of all the organisms and the water, SOIL and air surrounding them. The biosphere thus includes parts of the ATMOSPHERE, the LITHOSPHERE and most of the HYDROSPHERE. [J.G.]

biostratigraphic unit A biostratigraphical unit is defined on the basis of the fossil content or palaeontological character of a geological stratum. [M.O'C.]

biota The total flora and fauna of a region. [S.R.J.W.]

biotic factors Factors limiting the DISTRIBUTION of a species in space or time due to the effects of the animals, microorganisms and plants present. They are contrasted with the ABIOTIC FACTORS of climate, soil, wind, flood and fire. [P.J.G.]

biotic indices Biological criteria that are used to give information on ECOSYSTEM condition. Such indices have been used especially as indicators of riverwater quality. Use two observed characters of the effects of STRESS on communities: as POLLUTION stress increases the total number of species declines; moreover, as the extent of pollution stress increases, species tend to be selectively removed, with sensitive species (e.g. ephemeropterans) disappearing first and tolerant species (e.g. chironomids) last. [P.C.]

biotic integrity *See* INDEX OF BIOTIC INTEGRITY.

biotic potential *See* CARRYING CAPACITY.

biotope A region that is distinguished by particular environmental conditions and therefore a characteristic ASSEMBLAGE of organisms. *See also* HABITAT. [P.D.M.]

bioturbation Physical effect on SEDIMENT, by the activities of organisms in it. [P.C.]

biotype A specific type of plant or animal defined within a species; a group of individuals with similar GENOTYPES. *Cf.* BIOTOPE. [P.O.]

blanket bogs *See* BLANKET MIRE; MIRE.

blanket mire As its name implies, blanket mire is characterized by the widespread nature of the PEAT cover, often enveloping hill-tops, plateaux, slopes and valleys. [P.D.M.]

bloom A sudden increase in density of algae in a water body, usually due to an excess of NUTRIENTS. *See also* ALGAL BLOOM. [P.C.]

BOD *See* BIOCHEMICAL OXYGEN DEMAND.

body burden The total amount of XENOBIOTIC in an organism at a given time. [P.C.]

body size *See* ALLOMETRY; GROWTH; SIZE AT MATURITY.

bog A major subdivision in the classification of PEAT-forming ecosystems, or MIRES. It comprises those mires in which the supply of rainfall forms the sole input of water to the mire surface and which consequently receive their input of chemical elements only from dissolved materials or suspended dust (OMBROTROPHIC). [P.D.M.]

bomb calorimeter A device for measuring the heat of combustion of fuels or BIOMASS. [J.G.]

boreal A term meaning 'northern', 'of the North' (from *Boreas*, Greek and Latin, (God of) the north wind; *boreus* and *borealis* adj., Latin). [Y.V.]

boreal forest Boreal forest, or TAIGA, is the world's largest vegetation formation, comprising coniferous forests that stretch around the Northern hemisphere interrupted only by the North Atlantic Ocean and the Bering Strait. It forms a 1000–2000 km broad zone between treeless TUNDRA in the north and either broad-leaved DECIDUOUS FORESTS (in oceanic areas) or dry GRASSLANDS and semi-deserts (in continental areas) in the south. [Y.V.]

boring animals Animals that penetrate solid substrata by mechanical abrasion or chemical dissolution via the secretion of organic acids. [V.F.]

bottle effects Differences in growth or behaviour of aquatic organisms that are the result of being constrained in a relatively small volume where contact with the internal surface becomes important. [V.F.]

bottom-up controls Controls on the structure of a FOOD WEB operating through limitation of available RESOURCES, often NUTRIENTS and/or light in the case of plants or PRIMARY PRODUCTIVITY in the case of animals. It is contrasted with TOP-DOWN CONTROL, where predators at the top of the trophic hierarchy control web structure. The relative importance of top-down and bottom-up controls has been a source of considerable controversy, often with conflicting evidence for the same system. [S.J.H.]

boundary layer Water movements in RIVERS and streams are driven by gravitational gradients. The structure of this flow is mediated by friction induced by the channel boundary. The region where these frictional effects are felt is the boundary layer. [P.C.]

brackish habitats Brackish waters and soils contain ionic solutes at higher concentrations than those of FRESH WATER or normal soil solutions but at considerably lower concentration than seawater. They originate by dilution of seawater with rain or groundwater, or by evaporation of water and consequent enrichment with solutes in arid environments. [J.R.E.]

breeding value The breeding value, or additive genetic value, of an individual with respect to a QUANTITATIVE TRAIT measures the heritable effects of its GENOTYPE on its PHENOTYPE. [T.J.K.]

Brillouin index *See* DIVERSITY INDICES.

broken stick model The broken stick model was proposed by MacArthur (*see* CHARACTERS IN ECOLOGY) to predict the small-scale pattern of relative ABUNDANCE of species in an ASSEMBLAGE, assuming that species in a GUILD divide available RESOURCES among themselves and, in particular, that relative abundance is determined by one critical resource. MacArthur drew an analogy with a stick along which $n - 1$ points were picked at random where n is the number of species in the assemblage. When the stick is broken at each point the length of each segment corresponds to the abundance of one species. [M.H.]

brood parasitism The laying of EGGS in the nests of other individuals and the rearing of offspring by these conspecifics (intraspecific brood parasitism) or heterospecifics (interspecific brood parasitism). [A.P.M.]

bryophytes A distantly co-related group of non-vascular terrestrial plants comprising the mosses and liverworts. The bryophytes are of ancient LINEAGE, probably evolving directly from an algal ancestor at an early stage in land-plant EVOLUTION and possibly independent of the line which gave rise to vascular plants. Unlike vascular plants, the bryophytes have remained dependent on a close association with water for sexual reproduction. Unfortunately, because of their small size and relatively fine structure, bryophytes have not been readily recognized in the earliest FOSSIL RECORDS. The majority of contemporary species are still found most commonly in continuously moist habitats (though there are a number of exceptions with extraordinarily high capacities for tolerance of DESICCATION). The bryophytes are classified into three taxa: the mosses (Musci), with about 9000 described species, the liverworts (Hepaticae), with about 8000 species, and a small and questionably related group of hornworts (Anthocerotae), of about 250 species. [G.A.F.H.]

buffer zone A buffer zone is a boundary area surrounding or adjacent to a core conservation area, such as a park or nature reserve. [D.A.F.]

C

C:N ratio In soils and BIOMASS, a close relationship exists between the ratios of the elements carbon and nitrogen. The C:N ratio in arable soils is often from 8:1 to 15:1, with a tendency for the ratio to be lower in arid climates and lower in subsoils. In biomass, the C:N ratio reflects the content of CELLULOSE and lignin. For microorganisms the ratio is as low as 4:1 to 9:1, for leguminous plants it is 20:1 to 30:1, for straw residues it is 100:1 and for sawdust it is 400:1. Thus, the plant residues that become incorporated into the SOIL usually contain relatively little nitrogen in relation to the microbial population that they support. [J.G.]

C-value paradox The amount of DNA in the nucleus varies enormously between different species. The total amount of DNA in the haploid genome is known as the C-value. C-values range from less than 10^7 base pairs for Gram-positive bacteria, 10^8 for annelid worms, 10^9 for crustaceans and up to 10^{11} for angiosperms, with a trend towards increased amounts of DNA with increased complexity of the organism. The amount of DNA varies 5000-fold among algae, 1200-fold among pteridophytes and 2500-fold among angiosperms. The term 'C-value paradox' refers to the extraordinary variation in DNA amount and our inability to explain the function of the additional DNA, particularly in closely related species. [G.A.F.H. & I.O.]

calcicole A plant or organism limited to, or more abundant on, calcareous soils or in water of high calcium status (*cf.* CALCIFUGE). [J.R.E.]

calcifuge A plant or organism limited to, or more abundant on, soils or in water of low calcium status and usually of pH 5 or less (*cf.* CALCICOLE). [J.R.E.]

calorific value The calorific value of a material is a measure of its energy content expressed in JOULES per unit mass. Hence, it is probably more properly termed a 'joule equivalent'. It is most frequently determined by combustion in a BOMB CALORIMETER, which monitors the rise in TEMPERATURE when a known mass of dried material is ignited in an oxygen-enriched atmosphere. The instrument is calibrated with benzoic acid of which the calorific content is accurately known. Estimating the calorific values of materials enables us to express different components of FEEDING and PRODUCTIVITY equations in a common currency of energy units. *See also* ECOLOGICAL ENERGETICS. [M.H.]

cannibalism The killing and consumption of another member of the same species is widespread and often common. It has been reported in species as diverse as protozoans and mammals, and occurs for many different reasons. Adults may cannibalize their offspring to reduce COMPETITION between their young. Juveniles may eat competing siblings or unrelated eggs and juveniles. More rarely, adult females may attempt to kill and eat courting males, perhaps as a means of assessing male quality. [D.C. & M.A.E.]

canonical correlation A multivariate statistical method. [B.S.]

canopy The upper stratum of a WOODLAND or FOREST, comprising trees receiving full daylight over all or part of their crowns. [G.F.P.]

capital breeder An organism that uses stored energy for reproduction. The alternative is an INCOME BREEDER, that uses energy acquired during the reproductive episode. Examples of capital breeders include many species of temperate-zone fish and large mammals, which store energy in various body organs and then deplete these for the production of EGGS (fish) or to support lactation (mammals). *See also* COSTS OF REPRODUCTION; LIFE-HISTORY EVOLUTION. [D.A.R.]

captive breeding Breeding of animals, usually ENDANGERED SPECIES, in captivity, normally with a view to release back to nature and hence to the conservation of the species. [P.C.]

capture–recapture techniques When animals are marked, released and recaptured or otherwise re-examined, information may be obtained about their behaviour. For instance, distance moved may be estimated by recording displacement from release point to recapture point. A widely used application is for estimation of POPULATION SIZE. If a SAMPLE is taken, marked and released, marked and unmarked animals behave in the same way and the population is closed (i.e. with no input or loss), then the ratio of marked individuals to total population size is equal to the ratio of marked recaptures to total size of a second sample. For a population of unknown size P, if a sample n_1 is marked and released and there are m marked recaptures in a second sample n_2, then the population estimate is $P = n_1 n_2 / m$. It is not essential for the same effort to be put into capturing the two samples. The expression given above is commonly known as the LINCOLN INDEX or the Petersen estimate. [L.M.C.]

carbon cycle The carbon cycle is one of the four key BIOGEOCHEMICAL CYCLES on Earth: carbon, nitrogen, sulphur and phosphorus. All these elements are cycled around the globe between different components (also known as reservoirs or compartments) by physical, chemical and biological processes. Carbon in particular is a key element in life on Earth, being a component of an enormous range of organic compounds, formed because of the stable, long-chain covalent bonding possibilities of the carbon atom. Elemental carbon is rare (diamonds, graphite) and the commonest carbon compounds are the oxidized forms of carbon dioxide (CO_2) and carbonate. The organic carbon compounds have been reduced by the action of organisms, for example in PHOTOSYNTHESIS. Carbon is exchanged between the major reservoirs of carbon: the ATMOSPHERE, the OCEANS and FRESH WATER (although the latter is only a very small component), the terrestrial BIOSPHERE, and the sediments and sedimentary rocks. On very long GEOLOGICAL TIME-SCALES, sedimentary rocks are exposed, allowing weathering to return CO_2 to the atmosphere. However, the faster parts of the cycle are dominated by biological activity, removing carbon in the form of CO_2 from the atmosphere and from solution into living organisms (BIOMASS) during photosynthesis and releasing it during DECOMPOSITION (particularly the activity of microorganisms) as CO_2 and methane (CH_4). [J.I.L.M.]

carnivory This is applied to an organism that only consumes the tissue of other heterotrophic organisms. [R.C.]

carr A RHEOTROPHIC mire dominated by trees (often alders and willows). [P.D.M.]

carrion-feeders *See* FEEDING TYPES, CLASSIFICATION.

carrying capacity (*K*) This term describes a DYNAMIC EQUILIBRIUM around which a POPULATION fluctuates. Another analogous definition that is often used to describe this process is the equilibrium POPULATION DENSITY (N^*). This definition offers a more accurate reflection of natural populations as it does not imply that there is a maximum number of individuals that can be 'carried' in a particular habitat. The equilibrium will be such that when the population density is above it then the number of individuals will tend to decrease, and when the population density is below it then the number of individuals will tend to increase. [R.C.]

cascade model of food-web structure A MODEL to explain patterns observed in food-web datasets, so called because its central assumption is that species in the web can be ordered a priori into a FEEDING hierarchy. The model assumes that a species can never feed on those above it and that feeding links with those below it are chosen at random with a probability c/n, where c is a positive constant less than n (the number of species in the web). The number of links per species must be specified from available data before the model can be used. *See also* FOOD CHAIN; FOOD WEB. [S.J.H.]

caste The term 'caste' is used in biology to draw attention to differences among members of eusocial colonies. [N.R.F.]

castration, parasitic Infecting parasites cause HOST gonads not to function (e.g. by inhibiting development, causing loss once developed, causing malfunction). [P.C.]

casual species ALIEN species incapable of forming self-replacing populations and relying on continual reintroduction for persistence outside their native or NATURALIZED ranges. [M.J.C.]

catadromous A pattern of migration in which adults living in FRESH WATER migrate to breeding sites in saltwater. [P.O.]

catch per unit effort Return from fishing, hunting and SAMPLING. Usually follows law of diminishing returns; for example number of species sampled from a community increases but at a reducing rate as sampling effort is increased. Hence catch per unit effort varies with effort. [P.C.]

catchment A geographical region within which hydrological conditions are such that water becomes concentrated in a particular location, either a basin or a single RIVER by which the catchment is drained. The American term 'watershed' may be used as a synonym, but is employed rather differently in European literature where it implies a ridge separating catchments. [P.D.M.]

catotelm Lower functional SOIL layer in a MIRE: the PEAT deposit proper, perennially waterlogged in an intact system. From the Greek *kata*, down; *telma*, MARSH. The *inertnyy sloy* (inert layer) of Russian telmatologists. *See also* ACROTELM. [H.A.P.I.]

cellulase A complex enzyme system that hydrolyses CELLULOSE to sugars of lower relative molecular mass, including cellobiose and glucose. [P.C.]

cellulose Cellulose is classified as a polysaccharide comprising linked monosaccharide units. It is a straight-chain natural polymer consisting of glucose units with average relative molecular mass of more than 500 000. Cellulose forms the main structural units of all plants and is the most abundant polysaccharide occurring in nature. Cellulose is the main constituent of cell walls of all higher plants, many algae and some fungi. [M.C.G.]

censuses Estimates of POPULATION SIZES. [P.C.]

central limit theorem For large samples the sample arithmetic mean, $\bar{x}$, of n independent observations, taken from a POPULATION with mean μ and variance s^2, has approximately a NORMAL DISTRIBUTION with mean μ and variance s^2/n. This theorem applies to any PROBABILITY DISTRIBUTION with a well-defined mean and variance. This theorem is used to define the sample standard error

($\sqrt{s^2/n} = s/\sqrt{n}$), which is used in hypothesis testing, since it allows us to predict the probability of observing a sample estimate of $\bar{x}$ for a given population mean. Whilst the central limit theorem holds for large sample sizes, for smaller samples from normally distributed populations (approximately $n < 30$) it has been possible to predict the distribution of the sample mean relative to the population mean through Student's *t* distribution, which gives rise to the *t* TEST. *See also* STANDARD ERROR (OF THE MEAN). [R.P.F.]

centre of origin The area in which a taxonomic group originated and from which it dispersed. [P.O.]

centrifugal speciation *See* SPECIATION.

centrolecithal eggs *See* EGGS.

CFCs (chlorofluorocarbons) Synthetic substituted alkanes with one or more hydrogen atoms replaced by chlorine or fluorine, existing as gases or low-boiling liquids at normal temperatures and pressures. Chemically inert and non-toxic, CFCs are used as refrigerants, solvents, aerosol propellants and in blowing plastic foam, but are gradually being replaced by more reactive hydrochlorofluorocarbons (HCFCs) and hydrofluorocarbons (HFCs). They are important 'GREENHOUSE GASES', contributing about 10% to radiative forcing of the CLIMATE. CFCs accumulate in the TROPOSPHERE with lifetimes of tens to hundreds of years, and are transported to the STRATOSPHERE where they are photolysed and perturb the chemistry of ozone. The largest contributors are CFC-11 ($CFCl_3$), CFC-12 (CF_2Cl_2) and CFC-113 ($CF_2ClCFCl_2$). Production and use of CFCs and other halocarbons is now regulated under the 'Montreal Protocol', which seeks to prevent further depletion of stratospheric ozone and consequent increases in ultraviolet (UV)-B radiation at the Earth's surface. *See also* CLIMATE CHANGE; OZONE HOLE; OZONE LAYER. [J.N.C.]

chalk and limestone grassland Limestones are rocks composed mainly of calcium carbonate ($CaCO_3$), laid down in shallow seas. Chalk is a particularly pure form of limestone. The soils formed on these rocks are shallow, often with fragments of parent rock at or near the surface, and well drained. They are high in free calcium carbonate and have a high pH. They are also low in essential NUTRIENTS, especially nitrogen and phosphorus.

Chalk and limestone grassland is found in warm areas with low rainfall, often on steep slopes, and is maintained by GRAZING animals, mainly sheep and rabbits. Frequently no species becomes really dominant and there is often an abundance of broad-leaved herbs, many deep rooted. Disturbance caused by grazing animals creates bare sites that are often colonized by ANNUALS or short-lived PERENNIALS. [S.R.J.W.]

channelization The process of converting a meandering natural stream to a rip-rapped ditch, in order to reduce flooding of surrounding land by accelerating RUN-OFF. Channelization results in drastic reductions in the abundance of fish and invertebrates because habitat diversity is eliminated and water velocities are greatly increased during high-flow periods. [P.B.M.]

chaos Chaos is a term used to describe deterministic dynamic systems that are neither steady nor periodic, but which exhibit irregularity and complexity. Systems expressing chaotic dynamics display sensitive dependence on initial conditions. The sensitivity is exponential, meaning that as time goes on small errors in the solution (e.g. due to noise and computer round-off) grow exponentially. This means that short-term predictions may be made accurately, but it is not possible to make useful long-term predictions. Chaotic behaviour can be exhibited by deterministic equations and is called deterministic chaos. Chaotic behaviour exhibited by simple-looking equations was first described by the meteorologist Edward Lorenz in 1963. Since then, chaotic dynamics have been demonstrated in a wide variety of systems including fluids, plasmas, solid-state devices, circuits, lasers, mechanical devices, biology, chemistry, acoustics and celestial mechanics. An example of chaos in ecological systems is shown by the erratic, complex and unpredictable OSCILLATIONS in the size of populations having a high INTRINSIC RATE OF INCREASE (r). For example, if the net reproductive rate (R) in a discrete-time logistic population growth model is such that $3.57 < R < 4$ the population exhibits chaotic FLUCTUATIONS in size. *See also* POPULATION GENETICS. [V.F.]

chaparral shrublands Specifically, the term is applied to the fire-prone SCRUB vegetation of those areas of the western USA and north-western Mexico where Mediterranean-type climate (hot dry summers, mild moist winters) prevails. [R.M.C.]

character displacement A DISTRIBUTION pattern encountered in plant and animal populations where closely related sister species are recognizably different in zones of overlap (sympatric), but virtually indistinguishable where each occurs alone (allopatric). Two types are generally recognized. Ecological character displacement pertains to evolutionary change in the zone of overlap that results from INTERSPECIFIC COMPETITION for resources. Reproductive character displacement (REINFORCEMENT) involves changes in the prezygotic MATE RECOGNITION system that evolve as a response to selection against hybridization. [G.L.B.]

character polarity The plesiomorphic or apomorphic status of each TRAIT is called its polarity. The process of identifying the plesiomorphic and apomorphic traits is called character polarization. In PHYLOGENETIC SYSTEMATICS this is a logical process of assigning traits to plesiomorphic and apomorphic status based on a priori deductive arguments using outgroup comparisons. [D.R.B.]

characteristic return time A measure of the rate at which a population regains its equilibrium value after a perturbation. The characteristic return time is the inverse of the EXPONENTIAL GROWTH rate (r), i.e. the time taken for a population to increase in size by a factor of exp = 2.718 growing exponentially with rate constant r. Populations with a high r can rapidly track changes in the CARRYING CAPACITY or can recover from density-independent effects on the POPULATION GROWTH RATE. When the characteristic return time is less than the period of an environmental fluctuation divided by 2π, the population will track changes in the environment closely. *See also* DENSITY DEPENDENCE; POPULATION DYNAMICS; POPULATION REGULATION. [V.F.]

characters in ecology A list of some of the major characters of the past who have played some part in moulding ECOLOGY. They are ordered alphabetically.

Allee, W.C. (1885–1955). Studied forces determining social groupings (AGGREGATIONS) in animals.

Aristotle (384–322 BC). Greek philosopher. As well as forming the foundations of philosophy, is regarded by many as the father of natural history. His *Historia Animalium* contains much information on many species gained through observation and dissection.

Carson, R. (1907–1964). Author of the influential best-seller *Silent Spring* (1967) that highlighted the blight of the environment by PESTICIDES. So captured the public imagination that it launched the popular environment movement.

Clements, F.E. (1874–1926). American ecologist. Developed a holistic philosophy of ecology that represented the plant COMMUNITY as a SUPERORGANISM with a predictable development, i.e. SUCCESSION. Initiated detailed quadrat studies of plant communities.

Darwin, Charles (1809–1882). Originator of the concept of EVOLUTION by NATURAL SELECTION that was expounded in his book *On the Origin of Species* (1859) and in which the interaction, 'struggle', between organisms and environment 'for survival' is all important.

Darwin, Erasmus (1731–1802). Grandfather of Charles. Early theory of evolution based on a struggle for existence, but with impetus for change coming from the acquisition of characters for improvement. Hence more similar to Lamarckism than DARWINISM.

Dobzhansky, Th. (1900–1975). Pioneering work on POPULATION GENETICS using laboratory cultures of *Drosophila*.

Elton, C. (1900–1992). As Director of Bureau of Animal Populations in Oxford had enormous influence on development of British ecology. Launched long-term ecological study on Wytham Wood, Oxfordshire. His *Animal Ecology* (1972) is a classic and was the major textbook in ecology in Britain for many years.

Fisher, R.A. (1890–1962). One of the founders of POPULATION GENETICS. Also very influential in the application of rigorous statistical procedures in ecology.

Gause, G.F. (1910–1986). Performed experiments with protozoans to test predictions from Lotka–Volterra equations. Made explicit COMPETITIVE EXCLUSION principle, which is often known as the Gause principle.

Haeckel, E.H. (1834–1919). Originator of the term 'ecology' (Oecologie) in his *Generelle Morphologie* (1866) to denote interactions between organisms and external world. Derived from the Greek *oikos*, referring to operations of the family household.

Haldane, J.B.S. (1892–1964). With R.A. Fisher was one of the founders of POPULATION GENETICS. Drew attention to MELANISM in the peppered moth, *Biston betularia*, as an example of rapid evolution in action.

Hooker, J.D. (1817–1911). Succeeded his father William as Director of Kew Gardens. Collected plants on expeditions to the southern polar regions and to the Himalayas.

Hooker, W. (1785–1865). First official Director of Kew Gardens. Succeeded by his son Joseph.

Humboldt, Alexander von (1769–1859). Founder of botanical geography. Showed how physical environment determined DISTRIBUTION of plants on a geographical scale. Observed an equivalence between belts of vegetation ascending a mountain and geographical zones that girdle the Earth.

Hutchinson, G.E. (1903–1991). Promoted view that ecological relationships should be seen as systems governed by causal relationships. Raised a number of provocative issues for causal analysis: the n-dimensional NICHE; why there are so many kinds of species; ecology as the theatre in the evolutionary play.

Lack, D. (1910–1973). Tested predictions from evolutionary theory on field populations of birds. Worked with Galapagos finches, where he showed that very similar species had significant differences in feeding habits.

Lamarck, J.B. (1744–1829). Did important work on the CLASSIFICATION of animals. But best known for his theory of evolution involving the INHERITANCE OF ACQUIRED CHARACTERS.

Leopold, A. (1886–1948). In an influential essay, 'The Land Ethic', expressed the need for humanity to treat nature as something to which we belong, not as a commodity to be exploited.

Lindeman, R. (1915–1942). In a sadly short career, pioneered the view of ECOSYSTEMS as dynamic systems that involve flows of energy and cycles of matter.

Linnaeus, C. (1707–1778). Father of modern systematics. Author of *Systema Naturae* (1735). Initiated the binomial system of nomenclature.

Lotka, A.J. (1880–1944). Developed mathematical theory of POPULATION SIZE. Associated with V. Volterra.

Lyell, C. (1797–1875). Father of modern geology. Author of *Principles of Geology* (1830–1833).

MacArthur, R. (1930–1972). Student of G.E. Hutchinson. Expanded the Hutchinsonian approach and made it more rigorous and mathematical.

Nicholson, A.J. (1895–1969). Explored the different kinds of factors that might affect POPULATION SIZE, emphasizing the distinction between density-dependent and density-independent factors.

Shelford, V.E. (1877–1968). Applied Clements' concepts of COMMUNITY and SUCCESSION to animals.

Tansley, A.G. (1871–1955). Developed studies on plant communities in Britain following Clements' techniques but not his philosophy. Deeply opposed to notion of COMMUNITY as SUPERORGANISM.

Volterra, V. (1860–1940). Eminent Italian mathematical physicist. Developed mathematical theory of POPULATION SIZE. Associated with A.J. Lotka.

Wegener, A. (1880–1930). First champion of theory of CONTINENTAL DRIFT.

White, Gilbert (1720–1793). English naturalist. Author of *The Natural History of Selbourne*, a collection of letters written to fellow naturalists documenting observations on the WILDLIFE in his parish in southern England.

Wright, Sewall (1889–1988). Became interested in selection through animal breeding. Saw the importance of structure within populations to form subpopulations. Introduced concept of ADAPTIVE LANDSCAPE.

[P.C.]

cheating In the context of MUTUALISM, means not providing reciprocal benefits. [P.C.]

checkerboard distribution A species distribution pattern in which different species, with similar HABITAT requirements exclusively occupy interdigitated patches of habitat. [S.J.H.]

chemical oxygen demand (COD) Amount of oxygen consumed in the complete oxidation of carbonaceous material in a water sample as carried out in a standard test, usually using potassium dichromate as oxidizing agent. *See also* BIOCHEMICAL OXYGEN DEMAND. [P.C.]

chemolithotrophs These are anaerobic bacteria that use inorganic and organic compounds as sources of energy in the absence of light. [V.F.]

chi-squared test (χ^2) A non-parametric test that compares observed numbers, placed in categories, with those calculated (expected) on the basis of some hypothesis. It is used in two types of situation: GOODNESS-OF-FIT tests and CONTINGENCY TABLES. [B.S.]

choosiness Rejection of prospective mates. In most animal species females are more choosy than males. The magnitude and direction of this asymmetry arises from differences between the sexes in the effects of choosiness on reproductive output and survival. [J.D.R.]

chronic tests Testing for responses that are not immediately debilitating, such as impairment of growth, reproduction, propensity for diseases including cancer. Therefore these tests are carried out at relatively low concentrations over long time-spans and hence at low concentrations *Cf.* ACUTE TESTS. [P.C.]

chronospecies Also known as successional species, palaeospecies or evolutionary species, this is a palaeontological species concept used in the systematics of extinct organisms. When a LINEAGE reconstructed from fossils exhibits substantial evolutionary changes, it is convenient to divide its history into separate phases. The successive stages of the lineage, defined on the basis of morphological similarity, are called chronospecies. [P.O.]

circadian rhythms Endogenous self-sustained rhythm (OSCILLATION; sometimes called entrained rhythm) with a periodicity close to 24 h. Both the period and phase of the rhythm become entrained (synchronized) to 24 h by external cues (zeitgebers). [J.S.B.]

circumpolar Distributed around the regions of the North Pole (circumboreal) or South Pole (circumaustral or circumantiboreal). [R.M.M.C.]

cladism *See* PHYLOGENETIC SYSTEMATICS.

cladistic species concept The cladistic species concept is one of three major species concepts currently adopted by practitioners of phylogenetic systematics. It has two major elements. First, each species to be analysed phylogenetically must be recognized a priori by non-phylogenetic criteria, such as reproductive cohesion within the putative species or lack of interbreeding with other putative species. Second, each branch of a phylogenetic tree represents a distinct species. Terminal branches therefore represent species distinct from their non-terminal, or ancestral, antecedents, so no contemporaneous species are ever considered to be persistent ancestors. [D.R.B.]

cladogenesis Literally meaning branching divergent EVOLUTION, cladogenesis refers to the production of new species via the subdivision of an ancestral species. The term 'cladogenesis' encompasses an array of SPECIATION mechanisms, such as geographical subdivision of the ancestor or active movement of organisms followed by their subsequent isolation in a peripheral habitat. The end-result of such processes is the production of two or more species from one species. [D.R.B.]

cladogram A branching diagram of entities in which the branching is based on the inferred historical connections among the entities. These historical connections in turn are indicated by synapomorphies. Phylogeneticists thus tend to use the terms 'phylogenetic tree' and 'cladogram' interchangeably. Advocates of the derivative transformed cladistics movement, or cladists, have expanded the definition of cladogram to include not only the product of a phylogenetic systematic analysis (a phylogenetic tree) but also the results of phylogenetic studies in coevolution and historical BIOGEOGRAPHY (e.g. area cladograms or host cladograms). For some cladists, the term 'cladogram' is considered synonymous with any branching diagram of taxa derived in any manner (e.g. a phenogram). [D.R.B.]

classification The process of arranging a set of objects in an order. The procedure is usually hierarchical. In the biological context there are two objectives. One is to produce a system allowing different types of organism to be referred to unambiguously. The other is to arrange them in an order that corresponds as nearly as possible to their evolutionary histories. The two processes sometimes conflict. A standard nomenclature is now used, starting from species and grouping progressively into more inclusive categories of genus, family, order, class, phylum and kingdom. There are rules governing the use of these names, so that reference to particular types is unambiguous. [L.M.C.]

clay SOIL component consisting of mineral particles <2–4 μm, of low permeability and capable of being moulded when moist. [P.C.]

cleidoic *See* EGGS.

cleistogamy Cleistogamous flowers are those that never open, appearing not to develop beyond the bud stage. [I.O. & G.A.F.H.]

climate Climate is often regarded as 'average weather', i.e. a statistical description of the state of the ATMOSPHERE at any given location. The climate of an area can be described by its mean values of TEMPERATURE, rainfall, WIND speed or number of sunshine hours but it should also include some information on the extremes of the statistical distribution as it is often the extreme events that cause loss of life or economic hardship. [J.B.M.]

climate change Major changes in the Earth's energy balance and CLIMATE have occurred over geological time, and smaller changes are evident over the last few hundred years of recorded history. Climate change is not well understood as it involves many interacting factors, some of which are influenced by human activities but many others are natural. [J.G.]

climate zones The classification of the world into CLIMATE zones or regions, based partly on the mean annual rainfall and TEMPERATURE and also the soils and life forms of the VEGETATION. [J.G.]

climatology (i) The climatology of a place is a description of the long-term CLIMATE there; and (ii) climatology is also the scientific study of the climate, by which is meant the long-term patterns in the weather. [J.G.]

climax Climax was the name given by Clements (*see* CHARACTERS IN ECOLOGY) to the end-point of SUCCESSION. It has generally been defined as a kind of VEGETATION that is self-perpetuating in the absence of DISTURBANCE. [P.J.G.]

cline A GRADIENT in the genetic make-up of a species, typically along a spatial gradient in an environmental factor like altitude, latitude, salinity, photoperiod, etc. [M.J.C.]

clonal growth This type of GROWTH typically occurs in both plants and some lower invertebrates such as hydroids. It relies on basic cell division (by mitosis) to produce new genetically identical 'CLONES' of the original organism. Such 'modular' growth allows the organism to exploit a relatively large area without producing GAMETES and thus avoiding any parent–offspring COMPETITION for limited local RESOURCES. Vegetative growth in plants, particularly grasses, follows this pattern where a single genetic individual may occupy a relatively large area. This may consist of a single interdependent organism or a number of discrete subunits. [R.C.]

clone Genetically identical individuals, produced by mitotic cell division or physical division of viable plant parts (cuttings, rhizomes, grafted buds), which vary from one another only as a result of SOMATIC MUTATIONS. [M.J.C.]

closed community An ASSEMBLAGE of organisms covering the complete ground area of a habitat (*cf.* OPEN COMMUNITY), effectively precluding colonization by other species, as all ecological niches are already occupied. In closed vegetation, plants make continuous lateral contact. Closed communities are characteristic of late stages of SUCCESSION. [A.J.W.]

closed forest Stands of trees whose CANOPY cover is complete or nearly so. Individual tree crowns may not actually touch or overlap, but there are no gaps in the canopy into which a new tree may grow. Usually applied at a small scale, i.e. to stands of trees, but applicable to even-aged forests on a larger scale if there has been no recent thinning and no natural disturbances. [G.F.P.]

closed system A system is closed if there is no exchange of matter or energy with the outside world. [M.O.H.]

cloud forest A kind of TROPICAL MONTANE FOREST. Hot moisture-laden lowland air rises up mountain slopes. As it rises it cools and at a certain altitude a flat-bottomed cloud layer is formed where the dew point is reached. Clouds form and remain from mid-morning to evening, drastically reducing insolation. The regular presence of the cloud layer produces a strong ALTITUDINAL ZONATION, a sharp discontinuity in the forest vegetation.

The trees are closely spaced and there is a dense shrubby understorey in which tree-ferns are common. There are herbaceous vines, but woody climbers are rare. EPIPHYTES are abundant. [M.I.]

clutch size The number of offspring produced during a single breeding episode. [D.A.R.]

coarse-particulate organic matter (CPOM) Plant DETRITUS in aquatic systems >1 mm in diameter. In fresh water will be of the form of decomposing leaves, needles and woody material. *Cf.* FINE-PARTICULATE ORGANIC MATTER (FPOM) (<1 mm but >0.45 μm) and DISSOLVED ORGANIC MATTER (DOM) (<0.45 μm). [P.C.]

coastal Referring to processes or features of the shallow portion of the OCEAN, generally overlying the CONTINENTAL SHELF, where circulation and other features are strongly influenced by the bordering land. [V.F.]

coastal wetlands The mixing of fresh and saline waters in low-lying, flat coastal areas leads to the development of a series of wetland vegetation types, depending upon climate, water-flow patterns, sedimentation, etc. Fine-particle deposition in temperate regions leads to the development of SALT MARSHES, consisting largely of herbaceous and dwarf shrub vegetation. In tropical and SUBTROPICAL regions, arboreal vegetation dominates the SUCCESSION, forming mangroves. [P.D.M.]

cobweb graph This is a graphical method of analysis of POPULATION DYNAMICS that works well for organisms, such as ANNUAL plants, which have DISCRETE GENERATIONS. Using the NET RECRUITMENT CURVE it is possible to predict how many individuals there will be in the next generation. The result can be reflected about a line representing a stable population ($N_t = N_{t+1}$) and then the process is repeated allowing prediction of the POPULATION SIZE a further generation into the future. The technique gets its name because the resulting population trajectory may eventually resemble a web. [C.D.]

coefficient of community The relation of the number of species common to two COMMUNITIES to the total number of species in these communities. The Jaccard coefficient or index of similarity is often used, expressed as a percentage:

$$IS_J = \frac{a}{a+b+c} \times 100$$

where a is number of species common to both communities, b is number of species unique to one community and c is number of species unique to the other. Also frequently used is Sørensen's coefficient, which gives greater weighting to the species common to both samples than to those present in only one SAMPLE:

$$IS_S = \frac{2a}{a+b+c} \times 100$$

with notation as above. Both coefficients are used in calculations of similarity matrices for numerical CLASSIFICATION procedures. *See also* SIMILARITY COEFFICIENT. [A.J.W.]

coefficient of inbreeding Also known as the coefficient of kinship. This is the probability that two homologous genes, drawn randomly from the two individuals (one gene per individual), are identical by descent (i.e. are copies of the same gene in an earlier generation). [I.O.]

coefficient of variation The sample STANDARD DEVIATION (s) is an estimate of the variation in a POPULATION (σ). If we wanted to compare the variation in the mass of mice from several populations we could use the standard deviation. However, we could not compare the value of s for mass in mice and elephants, since the scale of the measurements are so different. One solution is to use the coefficient of variation, $CV = s/\bar{x}$, usually expressed as a percentage by multiplying by 100. [B.S.]

coevolution EVOLUTION in which the FITNESS of each GENOTYPE depends not only on the population densities and genetic composition of the species itself but also on the species with which it interacts. Coevolution has also been used to describe patterns of ADAPTATION exhibited in predator–prey, host–pathogen, competitive and mutualistic relationships. When the two species generate selective forces that direct the evolution of the other, the interaction between the species (whether it be antagonistic or cooperative) may continually escalate. [V.F.]

coexistence The permanent co-occurrence of two or more species in the same area. Mostly concerns potentially competing species. *See also* COEVOLUTION; COMPETITION; COMPETITIVE EXCLUSION. [J.G.S.]

cohort A group of conspecific individuals belonging to the same generation and therefore usually of approximately the same age. [A.J.D.]

cohort life table A LIFE TABLE constructed using data derived from following a single COHORT from the time at which the constituent individuals were 'born' to that at which the last representative dies. Also known as horizontal, dynamic or age-specific life tables. *See also* DEMOGRAPHY. [A.J.D.]

Cole's paradox The INTRINSIC RATE OF INCREASE of an ANNUAL species with a CLUTCH SIZE of c is equivalent to the intrinsic rate of increase of a PERENNIAL with a clutch size of $c+1$. As stated, the result is paradoxical because it suggests that the perennial habit should be very rare, which it is not. The result arises because L. Cole failed to consider differences in juvenile and adult survival: Cole's result only applies to the case in which adult and juvenile survival are equal and less than or equal to 1. The perennial habit is favoured by high adult survival and a low rate of increase. *See also* LIFE-HISTORY EVOLUTION; REPRODUCTIVE EFFORT. [D.A.R.]

coloniality The close spatial AGGREGATION of reproducing individuals of one or more species. [A.P.M.]

colonizing species All species are capable of some form

of DISPERSAL, but the phrase 'colonizing species' is used to describe species (often called RUDERAL or TRAMP SPECIES) that exhibit unusually effective long-distance dispersal (by wind, water or attached to, or in the guts of, animals). Colonizing species also exhibit the ability to establish with higher than average probability on arrival at their settling point, and are often associated with habitats created directly or indirectly by human activities (waste ground, arable fields, gardens, railway embankments, sea walls, etc.). [M.J.C.]

colony A POPULATION of organisms living in the same area. *See also* COLONIALITY; SOCIAL ORGANIZATION. [P.D.M.]

commensalism A situation in which two differing organisms characteristically and actively occur together, the one utilizing a food source provided mainly by the activities of the other, in which the utilizer therefore benefits from the association whilst the food provider is unaffected either positively or negatively. [R.S.K.B.]

commonness Can apply to a species that is locally abundant and/or has a wide geographical spread. *Cf.* RARITY, BIOLOGY OF. *See also* ABUNDANCE. [P.C.]

the Commons Major RESOURCES of the planet (ATMOSPHERE, water, SOIL) to which all have right of access and use, and which no one has a right to spoil. [P.C.]

communication The transfer of information between animals using visual, audible or chemical means. The term has been used of plants, which may release volatile chemicals into the atmosphere as a result of damage that can cause a response in neighbouring individuals. *See also* INFOCHEMICALS. [P.D.M.]

community The total living biotic component of an ECOSYSTEM, including plants, animals and microbes. The term (unlike 'assembly' or 'ASSEMBLAGE') implies interaction between the individuals and species in the form of COMPETITION, predation, MUTUALISM, COMMENSALISM, etc. [P.D.M.]

community assembly *See* COMMUNITY.

community continuum *See* CONTINUUM.

community ecology Community ecology is the study of the interactions between populations of organisms, and between populations and the physical ENVIRONMENT, in a particular community, and the effects that those interactions have on the behaviour and structure of that COMMUNITY. A community cannot simply be seen as the sum of its constituent species or populations, but has EMERGENT PROPERTIES that are not features of the component populations. The term, 'community' here encompasses a wide range of scales, and indeed one community can be described within another—thus a rotting tree stump may house a community of microorganisms, higher plants, and insects, whilst that stump might be part of a much larger community—a forest. *See also* ECOLOGY. [P.C.]

community matrix The matrix defining the strengths of all pair-wise species interactions in a COMMUNITY. In the case of a competitive interaction, for example, the strength of interaction would be given by the COMPETITION COEFFICIENT. Used in models of community dynamics. *See also* SPECIES INTERACTIONS. [J.G.S.]

comparative method This is, in fact, a whole battery of techniques designed to help make sense of cross-taxonomic DIVERSITY in character states and ecology. Two overriding patterns emerge from comparisons of TRAITS or characters across species. The first is that closely related species tend to be similar: birds have beaks and feathers and lay eggs, while placental mammals have teeth and fur and give birth to live young. The second pattern is that species with similar lifestyles also tend to be similar in ways that adapt them to their specific lifestyles: plants from different taxonomic families that live in deserts have fewer and smaller stomata. Because closely related species tend to live in similar environments and to have similar lifestyles, it is very easy to find a relationship between variation in two traits, neither of which is responsible for the other. For example, if we took no account of evolutionary affinity and we had a collection of birds and placental mammals, on finding that some species had beaks and others had teeth we might attempt to identify the selective forces in the environment responsible for the difference by examining differences in diet between beaked and toothed species in our sample. Modern comparative methods deal with this problem by using the principle of evolutionary convergence, whereby similar evolutionary responses occur in independently evolving lineages that have moved into similar environments. Each time the same character evolves in response to the same environmental change, we have an independent evolutionary event. For example, whenever female primates evolve PROMISCUITY, males respond to this by producing more sperm per ejaculate because this increases their chances of fertilizing receptive eggs (as with having more tickets in a lottery). As a consequence, relatively large testes size accords with female promiscuity in the family tree of primates. This means that comparative biologists need a good phylogenetic tree of the group being examined and need to locate with precision the evolutionary transitions on the branches of the tree. They can then examine whether evolutionary transitions between two characters or between a character and the environment that are thought to bear a functional relationship to each other tend to occur together more frequently than expected. It should also be borne in mind that biologists dealing with even the most sophisticated comparative methods are using arguments of inference based on correlational evidence; it is usually very difficult to separate cause from effect under such circumstances. [P.H.H.]

compartmentalization in communities Two extreme views of COMMUNITY organization are that: (i) every species interacts with every other; or (ii) species interact only with their food RESOURCES, specialist natural enemies and immediate competitors. In the second case, it is possible, in principle, to compartmentalize communities into a large number of relatively species-poor compartments. Interactions (FEEDING, COMPETITION, predation and PARASITISM) are predicted to be strong between members of the same compartment, but weak between members of different compartments. *See also* COMMUNITY ECOLOGY; SPECIES INTERACTIONS. [M.J.C.]

compensating density dependence Compensating DENSITY DEPENDENCE is where demographic parameters vary with POPULATION DENSITY in such a way that there is REGULATION of POPULATION SIZE via direct or negative density dependence. [R.H.S.]

compensation

1 Plant GROWTH: regrowth following tissue loss (e.g. from herbivory or physical damage), such that the final loss in reproductive performance is less than would be expected on the basis of the amount of tissue originally destroyed. It may involve redistribution of reserves (e.g. from root to shoot), production of epicormic shoots, increase in the duration of the growing period or increased net photosynthetic rates of surviving green tissues.

2 POPULATION DYNAMICS: a MORTALITY factor is said to be compensated when an increase in the death rate does not lead to reduced breeding POPULATION DENSITY (e.g. mortality caused by predation of juveniles may be compensated if adult population density is limited by the availability of nest sites). Generally, the later in life that a mortality factor operates and the less intense the DENSITY DEPENDENCE of subsequent mortality factors, the less likely it is that compensation will occur.

[M.J.C.]

compensation depth The depth, in an aquatic ecosystem, at which ATTENUATION of light limits gross photosynthetic production so that it is equal to respiratory carbon consumption; or, alternatively, the depth at which the amount of oxygen produced in PHOTOSYNTHESIS equals the oxygen consumed in RESPIRATION. [J.R.E. & V.F.]

competition Negative, i.e. detrimental, interaction between organisms caused by their need for a common resource. The effect of the interaction may be measured, for instance, by the change in equilibrium POPULATION DENSITY or by its consequences for the (absolute) FITNESS of the individuals involved. Competition may occur between individuals of the same species (INTRASPECIFIC COMPETITION) or of different species (INTERSPECIFIC COMPETITION). [J.G.S.]

competition coefficient The relative per capita effect (on population growth) exerted by a competing species; in other words, the number of conspecifics needed to cause the same competitive effect as one individual of the competing species. [J.G.S.]

competitive exclusion The (local) extinction of a species due to INTERSPECIFIC COMPETITION. The 'principle of competitive exclusion' (also called Gause's principle; *see* CHARACTERS IN ECOLOGY) states that species cannot coexist (*see* COEXISTENCE) as long as they occupy the same NICHE or, in more precise terms, as long as they are limited by the same RESOURCE. The theory that led to the principle of competitive exclusion assumes a homogeneous environment, i.e. an environment without temporal changes or spatial structure. The ubiquitous variation in natural systems therefore severely limits the applicability of the principle. *See also* LIMITING SIMILARITY. [J.G.S.]

competitive release The expansion of the NICHE following the removal of a competing species. [J.G.S.]

complex life cycles A LIFE CYCLE of an organism is considered to be complex if at least two of the phenotypes produced in the cycle differ dinstinctly in morphology. Complex life cycles (CLCs) are very prevalent among animals. The most common form of CLC is that of a larval stage transforming into a sexually mature adult form via METAMORPHOSIS. [W.W.W.]

complexity of ecosystems Strictly speaking, the term 'ECOSYSTEM complexity' should denote some integrated measure of the complexity of the biological COMMUNITY, together with its physical ENVIRONMENT. In practice, however, it is often taken to refer only to the biological community, the complexity of which often reflects that of the physical environment (*see* HABITAT STRUCTURE). The mean number of species in the system, the number of interactions between species or some combination of these and related measures can be taken as indices of complexity. [S.J.H.]

complexity–stability The complexity of a COMMUNITY is measured as the number of species, the number of interactions between species, the average strength of interaction, or some combination of these. Stability is a measure of a community's sensitivity to DISTURBANCE. Stability may be measured in terms of the number and ABUNDANCE of its component species or by other community properties or processes. The concept of stability includes the property of RESILIENCE (the speed with which a community returns to its former state after it has been displaced from that state by a perturbation) and RESISTANCE (the ability of a community to avoid displacement in the first place). Communities may be described as locally stable (able to recover from small perturbations) or globally stable (able to recover from large perturbations). If a community is stable only within a narrow range of environmental or biological conditions it is said to be DYNAMICALLY FRAGILE; if it is stable over a broad range of conditions it is described as DYNAMICALLY ROBUST. During the 1950s and 1960s, ecologists believed that increased complexity within a community would lead to increased stability. More recent models of communities suggest that stability tends to decrease as complexity increases. However, evidence from real communities indicates that the relationship between community complexity and stability varies with the precise nature of the community, with the way in which the community is perturbed and with the way in which stability is assessed. [V.F.]

condition index (CI) Index of metabolic condition of an individual, usually in terms of mass per unit length. [P.C.]

confidence limit Confidence limits describe where we would expect a population parameter to lie in relation to a statistic estimated from a SAMPLE. For example, since the SAMPLING distribution of means (from samples of size $n \geq 30$) (*see* STANDARD ERROR) is a NORMAL DISTRIBUTION, we know that 95% of observed means will be within the interval $\mu \pm 1.96s/\sqrt{n}$, where $s/\sqrt{n}$ is the estimated standard error of the MEAN. Or, stated the other way round, we can be 95% confident that the interval $\bar{x} \pm 1.96s/\sqrt{n}$ will contain the population mean (μ). This interval is known as a confidence interval or, more precisely, as the 95% confidence interval. The interval $\bar{x} \pm 2.58s/\sqrt{n}$ is known as the 99% confidence interval. Clearly, if the confidence interval is large we can place less reliability on the sample mean as an estimate of the population mean. [B.S.]

conflicts A conflict may be said to exist between two parties if an action by the first party increases the FITNESS of that party (and hence the gene for the TRAIT tends to spread in a population) and decreases the absolute fitness of a second party (which is not in direct COMPETITION with the first) such that spread of trait 1 creates the context for the spread of a suppressor of the action performed by the second party. [L.D.H.]

conformer An organism whose physiological state (e.g. body temperature (thermoconformer) or body fluid composition (osmoconformer)) is identical to, and varies identically with, that of the external environment. Some species may show conformance with respect to some environmental variables but REGULATION with respect to others. [V.F.]

connectance Fraction of all possible pairs of species within a COMMUNITY that interact directly as feeder and food. In other words, the number of actual connections in a FOOD WEB divided by the total number of possible connections. *See also* TROPHIC CLASSIFICATION. [P.C.]

conservation biology Conservation biology is an emerging, interdisciplinary field that seeks to establish a scientific basis for the conservation and management of populations, communities and ECOSYSTEMS. At the same time, conservation biology draws on the empirical observations and results of land management practices as a primary source of information and insight. Consequently, conservation biology may be thought of as the interface between ECOLOGY and allied disciplines on one hand, and the practice of conservation management on the other. [D.A.F.]

constant-effort harvesting This type of HARVESTING is theoretically attractive as it involves natural FEEDBACK between the level of catch and the population of the harvested resource. The idea is that the amount of MORTALITY inflicted remains effectively constant. This can be done by setting, for example, a restriction on the amount of hunters or, in the case of fishing, the number of fishing vessels of a certain power. The operation of this constant amount of effort means that the catch will vary in proportion to the POPULATION SIZE. This makes an implicit assumption that catch is related to effort and population size by a simple linear relationship. The constant of proportionality is usually termed in fisheries as the catchability coefficient. The basic equation is:

$$\text{Catch} = \text{Catchability coefficient} \times \text{Effort} \times \text{Population size}$$

In theory this is an ideal mechanism, as when the population is small the catches are small. However, there are problems in its practical application as technological improvements occur and the effective effort can increase undetected. Furthermore, in certain cases, particularly involving RESOURCES that aggregate spatially, the relationship between catch per unit of effort and population size is non-linear, which can mask overexploitation. *See also* HARVESTING; SUSTAINABLE YIELD. [J.B.]

constant final yield This involves the idea of a population equilibrium in which the removals by HARVESTING (the YIELD) exactly equal the surplus that comes from the balance of births, deaths and growth in the population. [J.B.]

constraints In its formal mathematical context, constraint refers to any internal factor that interferes with the optimizing function of NATURAL SELECTION by preventing the required variation to be expressed in the population under consideration. [P.A.]

constraints on foraging Optimality (*see* BEHAVIOURAL ECOLOGY) approaches to foraging behaviour have seen the advent of rigorous, quantitative models that generate predictive hypotheses about foraging decisions (e.g. where to start foraging, how long to stay there, what sort of prey to take, etc.). Like all OPTIMALITY MODELS, optimal foraging models make a number of basic assumptions, for instance about what selection has shaped a forager to do (e.g. maximize its net rate of energy intake, minimize the risk of doing badly) and about the reproductive costs and benefits accruing from alternative foraging strategies. One important set of assumptions concerns the CONSTRAINTS that limit the predator's options at any given decision point. Some constraints are imposed by the environment. For example, the rate at which a bumble-bee can extract nectar from a flower may be limited by the morphology of the flower and the viscosity of the nectar. The risk of predation may be another limiting factor if, say, good food sources tend to occur in exposed areas where the forager is vulnerable to attack. Other constraints may be inherent to the forager itself. Thus perceptual limitations may constrain the ability to distinguish between certain kinds of prey, or wing-loading characteristics may set a limit to the amount of food that can be harvested before flying back to the nest. A troublesome possibility is that the absence of appropriate genetic variation in the forager's evolutionary past may act as a constraint on its present options. The judicious incorporation of constraints into foraging models is clearly a crucial prerequisite to testing the models fairly. However, it is important that convenient 'constraints' are not used uncritically to improve the match between prediction and empirical outcome. *See also* FORAGING, ECONOMICS OF; TROPHIC CLASSIFICATION. [C.J.B.]

consumer organisms Consumer organisms are all heterotrophic organisms that ingest (consume) other live plants or animals or parts of other live plants or animals and so include herbivores, carnivores and omnivores. [M.H.]

consumption Consumption is that part of the material removed by any trophic unit: individual, population, group of populations or trophic group, or even whole TROPHIC LEVEL, from a lower trophic level that passes into the body of the organism(s). [M.H.]

contagious distribution *See* AGGREGATED DISTRIBUTION.

contaminant A non-natural substance in the natural environment, not necessarily causing HARM. [P.C.]

contest competition A term originally coined by A.J. Nicholson in 1954 (*see* CHARACTERS IN ECOLOGY) to describe resource use, though frequently used now to describe both the behavioural process and the ecological outcome of INTRASPECIFIC COMPETITION. In contest competition the individuals may be said to compete for prizes. They are either fully successsful, or unsuccessful. The whole amount of the resource obtained collectively by the animals is used effectively and without wastage in maintaining the population. Contest competition often involves some form of direct INTERFERENCE, for example aggressive interactions between male birds competing for territories. In pure contest competition, the number of 'winners' (e.g. territory holders or survivors) is constant. The ecological outcome of pure contest competition is that there is EXACT COMPENSATION and POPULATION DYNAMICS tend to exhibit dynamic stability, in contrast with SCRAMBLE COMPETITION. *See also* ASYMMETRIC COMPETITION. [R.H.S.]

continental drift The lateral mobility of continents over geological time. [A.H.]

continental shelf A broad expanse of ocean bottom, representing the submerged edge of a continent. [V.F.]

continental slope The relatively steep downward slope extending seaward from the outer edge of the CONTINENTAL SHELF to the flat ocean floor. The continental slope usually extends to a depth of 2000–3000 m and varies in width between 20 and 100 km. It is usually covered by sediments of fine silt and mud. [V.F.]

contingency table Most frequently a two-way table (at least two rows and two columns) in which qualitative information is displayed prior to analysis for ASSOCIATION. *See also* CHI-SQUARED TEST. [B.S.]

continuum Mathematically, a continuum is an infinite set of objects between any two of which a third can always be interposed. The real numbers form a continuum, whereas the integers are a discrete set. The mathematical concept has had a powerful influence on ecology, providing a paradigm according to which communities are seen as nature's response to continuous underlying environmental variation. [M.O.H.]

convergence of successions According to F.E. Clements's (*see* CHARACTERS IN ECOLOGY) original theory of CLIMAX, the VEGETATION on all sites under a given climate should converge with time on a single species or single group of species, even where the early successional species are quite different, for example in lakes, on rocks or on SOILS with different chemical properties, etc. Some studies have shown evidence for convergence but, in general, critical studies have led to the rejection of this viewpoint and replacement with the idea of polyclimax. *See also* SUCCESSION. [P.J.G.]

convergent evolution Appearance of similar PHENOTYPES or GENOTYPES as a result of similar SELECTION PRESSURES rather than as a result of common ancestry (HOMOLOGY). [D.A.B. & D.R.B.]

conveyor-belt species Species of deposit-feeding animals that live infaunally (i.e. buried within the SEDIMENT) and that orient with their head down and their tail end at the sediment surface. Because they feed on sediment at depth and defecate at the sediment surface, they act as geochemical conveyor belts, redistributing sediment and altering its physical and geochemical properties in a predictable fashion. *See also* ECOSYSTEM ENGINEERS. [V.F.]

cooperative breeding, evolution of Cooperative breeding (or 'helping at the nest') occurs when individuals exhibit parent-like behaviour toward young that are not their own. Most commonly, cooperative breeding involves non-breeding individuals called 'helpers', 'auxiliaries' or 'workers' that forgo reproduction of their own while aiding in the reproductive efforts of others. [W.D.K.]

Cope's rule The observation that mammals and other lineages tend to increase in SIZE over evolutionary time. The explanation for this pattern is thought to be that smaller species survive MASS EXTINCTIONS better than large species. The ADAPTIVE RADIATIONS following an extinction event tend to involve evolutionary increases in size to fill newly vacated ecological niches. [D.A.B.]

coprophagy A specialization in FEEDING on faeces. *See also* FEEDING TYPES, CLASSIFICATION. [P.O.]

coral reefs Coral reefs are named for that group of cnidarians (belonging to the class Anthozoa) that secrete an external skeleton of calcium carbonate. They are found in shallow waters surrounding tropical landmasses. They are restricted to high SALINITY, silt-free waters warmer than 18°C, generally in a band that lies between the Tropic of Cancer and the Tropic of Capricorn. There are three types of coral reefs: atolls, barrier reefs and fringing reefs. [V.F.]

core species These are species likely to be regionally common, locally abundant and relatively well spaced in NICHE space. Such species are very likely to remain in a COMMUNITY despite a general TURNOVER of species in the system. [C.D.]

correlation When two sets of measurements (often called x and y) are related or show ASSOCIATION they are said to be correlated. If one (y) increases when the other (x) increases they are said to be positively correlated. If one decreases when the other increases they are said to be negatively correlated. Both parametric (PRODUCT–MOMENT CORRELATION COEFFICIENT) and non-parametric (KENDALL'S RANK CORRELATION COEFFICIENT and SPEARMAN'S RANK CORRELATION COEFFICIENT) statistics are available. Remember that a correlation between two variables does not necessarily mean that one causes the other. It may be that the two variables are both effects of a common cause. [B.S.]

corridor A corridor is a linear patch of HABITAT, usually established or maintained to connect two or more adjacent habitat areas. Examples of corridors include HEDGEROWS, railroad and highway rights-of-way, forested shelter-belts, and GALLERY FORESTS along RIPARIAN zones. [D.A.F.]

corridor dispersal When formerly isolated landmasses become linked by the emergence of an isthmus, terrestrial organisms can intermigrate across the newly created CORRIDOR. [A.H.]

cosmopolitan distribution Taxonomic groups that occur throughout the world in suitable habitats. [A.M.M.]

cosms Artificial ECOSYSTEMS of various levels of complexity and SCALE. *See also* MACROCOSM; MESOCOSM; MICROCOSM. [P.C.]

cospeciation Studies of cospeciation attempt to uncover the patterns of geographical or ecological associations among clades that share a close and evident ecological association (such as host–parasite, predator–prey, herbivore–crop or host plant–pollinator). There are two components to these patterns. First, two or more species may be associated today because their ancestors were associated with each other in the past (association by descent). Second, two or more species may be associated today because at least one of them originated in some other context and subsequently became involved in the interaction by colonization of a host or DISPERSAL into a geographical area (association by colonization; dispersal or host-switching). [D.R.B.]

costs of reproduction Deleterious effects on survival or FECUNDITY that occur as a consequence of current REPRODUCTIVE EFFORT. [D.A.B.]

costs of resistance Resisting adverse effects from outside agents is often metabolically costly: escaping a predator requires increased activity; resisting parasitism or disease involves the deployment of active defence mechanisms; resisting chemical poisoning may require avoidance behaviour, active transport and excretion processes, detoxification and repair of damaged tissues. These can all be considered as costs of RESISTANCE. It is presumed that some costs trade-off with other elements of FITNESS. [P.C.]

costs of sex Although nearly all organisms have some form of sexuality, the widespread existence of sex has been considered paradoxical because of the measurable costs involved with this mode of reproduction. Although evolutionary advantages for the population and species may result from sexual recombination (and the resulting creation of genetic DIVERSITY), advantages to the individual are less clear. The costs of sex include the time and energy used to find a mate, to engage in COURTSHIP and to make and maintain secondary sexual organs (e.g. organs for sexual display). In addition, more energy is generally invested in the production of EGGS compared with sperm, and frequently the female invests more in her offspring than does the male. There is therefore a cost to the female of investing in offspring that only carry half of her genes. This is sometimes referred to as the cost of producing males. The cost of sex in dioecious populations depends on the sex ratio, the advantage of an asexual CLONE becoming two-fold when the proportion of males in the sexual population is 50%. The cost of male allocation (less efficient production of offspring) can be distinguished from the cost of meiosis (reducing the genetic contribution to one's offspring) and which of these costs is more important may vary depending on specific circumstances. A reduction in genetic FITNESS due to reassortment includes losses due to both segregation and recombination, and may put sexuals at a disadvantage compared to asexuals in which superior epistatic gene combinations are preserved. However, the increased FECUNDITY inherent in asexuality may be offset by the wider range of RESOURCES available to the more genetically heterogeneous sexuals. Although the maintenance of sex has long been one of the most hotly debated issues in evolutionary biology, the explanation for the relative preponderance of sexual reproduction has not been fully resolved. *See also* EVOLUTIONARY OPTIMIZATION; LIFE-HISTORY EVOLUTION. [V.F.]

courtship Behavioural patterns preceding copulation or the formation of social bonds between individuals of the opposite sex. [A.P.M.]

covariance An analogous term to variance. While the variance is a measure of how single observations (*x*) in a SAMPLE vary, the covariance is a measure of how pairs of observations (*x* and *y*) in a sample covary (i.e. the degree of ASSOCIATION between them). It contributes to the PRODUCT–MOMENT CORRELATION COEFFICIENT. [B.S.]

cover The proportion of ground occupied by the perpendicular projection on to it of the aerial parts of individuals of each species present. [A.J.W.]

critical loads Quantitative estimate of exposure to one or more pollutants below which significant harmful effects on specified sensitive elements of the environment do not occur according to present knowledge. [P.C.]

crop Species used in agriculture and their YIELD; usually refers to plants (crop of maize, of rice; good crop from a harvest) but occasionally to animals (e.g. output/yield of fish farm in AQUACULTURE). [P.C.]

crust, Earth The solid outer portion of the Earth. Composed primarily of oxygen (46.60% by mass), silicon (27.72%), aluminium (8.13%), iron (5.00%), calcium (3.63%), sodium (2.83%), potassium (2.59%), magnesium (2.09%) and titanium (0.44%), which make up various silicate minerals. Crustal thickness and elastic properties vary widely and abruptly. Continental crust is lighter (average density = $2.7\,g\,cm^{-3}$) and thicker (average thickness = 50 km) than oceanic crust (average density = $3.0\,g\,cm^{-3}$; average thickness = 7 km). [V.F.]

cryopreservation The keeping of organisms, or their propagules, at low temperatures and hence extending their normal lifespans, usually for purposes of selective breeding and conservation. [P.C.]

crypsis A STRATEGY adopted by prey to reduce the probability of detection by a PREDATOR by increasing their resemblance to a random sample of the environment. [D.A.B.]

cryptobiont Literally, a hidden organism. Usually applied to dormant spores (often of microbes) within an ecosystem that can remain viable but difficult to detect for considerable periods. [P.D.M.]

cryptogamic soil crust Also called microbiotic SOIL crusts, these are delicate symbioses of cyanobacteria, lichens and mosses from a variety of taxa that inhabit arid and semi-arid ecosystems. Crusts perform several essential ecological functions: increase organic matter and available phosphorus, soil stability and water infiltration; provide favourable sites for the germination of vascular plants; and, most crucially, perform the major share of nitrogen fixation in DESERT ecosystems. [T.L.F.]

cryptophyte A plant life form in which the organ of perennation is hidden in SOIL or mud during a period unfavourable for GROWTH. It includes geophytes, where survival depends upon below-ground protection (e.g. plants with bulbs, corms, tubers, etc.), and HYDROPHYTES, where the perennating organs are protected by submersion. *See also* PHYSIOGNOMY. [P.D.M.]

culling This term tends to be applied to the HARVESTING of large mammals, including marine mammals. [J.B.]

$CV^2 > 1$ rule In many discrete-time MODELS OF PARASITOID–HOST INTERACTIONS derived from the Nicholson–Bailey model, stability requires that HOSTS vary in their risk of parasitoid attack and that the square of the coefficient of variance of risk among hosts is greater than unity. [H.C.J.G.]

cyclic succession Cyclic SUCCESSION occurs where a given species or group of species is locally destroyed and there is succession back to that or those species. [P.J.G.]

cyclomorphosis Cyclic change in PHENOTYPE, such as seasonal changes in morphology, particularly conspicuous among cladoceran Crustacea and rotifers (phylum Rotifera), also described in dinoflagellates and protozoa. Its expression is based on environmentally induced (e.g. daylength, temperature, presence of predators) PHENOTYPIC PLASTICITY. [D.E.]

D

Darwinism The theory of EVOLUTION by NATURAL SELECTION as elaborated by Charles Darwin (1809–1882) (*see* CHARACTERS IN ECOLOGY). [L.M.C.]

data logger A programmable electronic device for capturing data. [J.G.]

day-degrees Product of the number of days and number of degrees (°C) by which the TEMPERATURE exceeds an arbitrary or predetermined THRESHOLD. Day-degrees (DD) is a cumulative index that quantifies the thermal budget required for an organism to develop from one growth stage to another. [J.S.B.]

day number Day number is a chronology describing the number of days since the first day of January within a year: 1 January is defined as having a day number of 1. Day number is frequently confused with JULIAN DAY, which counts the number of days since noon on 1 January 4713 BC. *See also* TIME. [P.R.V.G.]

daylength The period of daylight between sunrise and sunset, which are defined as the times at which the true position of the centre of the solar disc passes over the horizon. [P.R.V.G.]

DDT (dichlorodiphenyl-trichloroethane) DDT is a potent stomach and contact organochlorine INSECTICIDE, discovered in 1939, and considered a revolutionary development in PEST CONTROL. It was apparently much safer to humans than earlier insecticides and was widely used to control human lice and disease-transmitting insects in the Second World War. DDT was also used on a wide range of crops and on storage pests. However, it has been widely banned because of environmental persistence and BIOACCUMULATION in animal body fats and the FOOD CHAIN, with concerns regarding reproduction in higher animals, notably thinning of egg-shells of raptorial birds. *See also* PESTICIDE. [K.D.]

de Wit replacement series An experimental protocol for studying interspecific plant COMPETITION, named after the Dutch ecologist and agronomist C.T. de Wit; also known as substitutive experiments. The technique involves holding total plant DENSITY constant and varying the proportion of two competing species. The technique is used for the computation of indices of competition. Replacement series have been criticized for their assumption that total plant density (A plus B) is constant and that the choice of this constant density is arbitrary. An ideal design would alter both density and frequency in a factorial combination. *See also* ADDITIVE EXPERIMENTS. [M.J.C.]

dead zones Zones of reduced flow that retain water in a flowing-water body. If the bottom is 'rough', there will be many such zones, chiefly on the downstream side of large boulders where flow is reduced. The retained water in these zones is nevertheless progressively renewed by fluid exchange with the main flow. These areas can support an extensive PLANKTON community. They also act as REFUGES for organisms and as temporary sinks for sediments under high flows. *See also* BOUNDARY LAYER. [P.C.]

deciduous forest Forest whose trees reduce water loss by dropping their leaves during the unfavourable season of the year, which is winter in the case of summer-green deciduous forests. [J.R.P.]

decomposers *See* DECOMPOSITION.

decomposition Breakdown of chemicals to simpler products, i.e. nearer thermodynamic equilibrium. Often applied to breakdown of BIOMASS under physical and biological action. [P.C.]

decreaser A plant that declines in ABUNDANCE on heavily grazed PASTURE or rangeland. Decreasers tend to be palatable species, often more competitive than increasers in the absence of herbivory (this implies the existence of a TRADE-OFF between competitive ability and palatability, in which there are competing demands within the plant for investment in GROWTH versus defence). [M.J.C.]

deep sea The part of the marine environment that lies below the level of effective light penetration for PHYTOPLANKTON photosynthesis in the open OCEAN and below the depth of the continental shelves (greater than *c.* 200 m). It is sometimes referred to as the APHOTIC ZONE. [V.F.]

defence mechanisms Organisms are subject to a variety of threats from MORTALITY agents: some are associated with external (extrinsic/ecological) agents such as accidents, predators, diseases and toxic chemicals; and some are due to internal factors associated with system deterioration or AGEING. Against each, organisms can deploy a variety of responses, collectively referred to as defence mechanisms. For threats from predators, they can involve avoidance and disuasions or repulsion. For diseases, they can involve exclusion of the disease-causing organisms and/or the deployment of immunological systems. For pollutants, they can involve avoidance, exclusion, excretion or neutralization (e.g. by induction of protective proteins). These toxicants, and the processes that lead to ageing, can cause damage to tissues and macromolecules, so another form of defence is by repair processes which, within organisms, invariably involves cellular and molecular turnover. [P.C.]

defoliant Substance causing loss of leaves from plants. *See also* BIOCIDE; HERBICIDE. [P.C.]

deforestation The removal of CLOSED FOREST or open WOODLAND. It has been taking place all over the world for millennia. Commonly, deforestation is to create land for agriculture; if this is abandoned, shrublands and tree COVER eventually re-establish unless the site has become very seriously eroded. [T.C.W.]

degrees of freedom (υ) From a purely practical point of view this is a number that we need to calculate, in most statistical tests, before we can look up the significance of the test statistic in the appropriate STATISTICAL TABLE. For example, in the t TEST it is $n_1 + n_2 - 2$ and in the chi-squared GOODNESS-OF-FIT test it is frequently $n - 1$ (where n is the number of categories). However, this term also refers to the number of independent items of information in an experiment or statistical test. [B.S.]

delayed density dependence A change in POPULATION GROWTH RATE associated with a change in POPULATION DENSITY some time previously. [R.H.S.]

delayed semelparity Life-history pattern of long-lived organisms that reproduce only once. Examples include some salmon, yucca, lobelias and cicadas. [D.A.B.]

deme A local, randomly mating POPULATION, partially isolated from other such local populations; a basic unit of genetic population structure. *See also* LINEAGE; QUANTITATIVE GENETICS. [T.J.K.]

demersal Living on or near the bottom of a SEA or LAKE. *See also* BENTHIC HABITAT CLASSIFICATION. [P.O.]

demographic stochasticity The chance variation in the number of integer births and deaths in a population with time-invariant vital rates. [S.J.H.]

demography The processes of birth, death, IMMIGRATION and EMIGRATION that determine the size, FLUCTUATIONS and AGE STRUCTURE of populations. Also the study of these processes and their effects. *See also* LIFE TABLE; POPULATION DYNAMICS. [A.J.D.]

dendrochronology The use of data on the widths of annual tree rings for the purposes of dating past events. [M.J.C.]

dendrogram A tree-like diagram designed to show postulated phylogenetic relationships between taxa. [P.O.]

denitrification The microbial conversion of nitrate (NO_3^-) ions into nitrogen gas occurs mainly in anaerobic soils, especially at high pH, fresh water and seawater. In conditions where DISSOLVED OXYGEN is unavailable, certain bacteria (e.g. some species of *Bacillus* and *Pseudomonas*) are able to use nitrate ions in RESPIRATION during the DECOMPOSITION of organic materials within the SOIL. In the process, nitrate is reduced to nitrogen gas or nitrous oxide. Approximately 180 million tonnes of atmospheric nitrogen are generated globally per year in this way, which is roughly equivalent to the total biological fixation of nitrogen. *See also* NITROGEN CYCLE. [P.D.M.]

density The number of individuals per unit area (terrestrial species) or per unit volume (aquatic or aerial species). Density is often used to quantify the quality or attractiveness of a HABITAT or a patch of habitat, for example prey density is a measure of food availability, while tree density might help to predict the attractiveness of an area for a tree-dwelling species. [R.H.S.]

density dependence Density dependence is where one or more demographic parameters (birth, death, IMMIGRATION or EMIGRATION rates) is a function of POPULATION DENSITY, i.e. a change in population density allows prediction of a change in a demographic parameter. [R.H.S.]

density-dependent selection Density-dependent selection implies that the FITNESS differences underlying selection in the population depend upon the total number of individuals of a specified AGE CLASS or life stage. *See also* NATURAL SELECTION. [G.D.J.]

density independence Density independence is the absence of a relationship between any demographic parameter and POPULATION DENSITY. [R.H.S.]

density vagueness A concept pertaining to REGULATION of populations whereby density-dependent regulation is only manifest at very high (and possibly very low) densities, with populations fluctuating at random or in response to variation in the environment at intermediate densities. [S.J.H.]

deposit-feeder An aquatic animal that feeds on the surface layer of SEDIMENT, for example certain polychaete worms. *See also* FEEDING TYPES, CLASSIFICATION. [D.A.B.]

desert Approximately 45% of the Earth's land surface is occupied by desert (which may be hot or cold), characterized by dry conditions and low BIOMASS and plant PRODUCTIVITY. Despite having a PRIMARY PRODUCTIVITY rate of only about 0.3 kg m^{-2} $year^{-1}$, the deserts support about 13% of the world's human population.

Defining deserts in terms of precipitation is difficult because the TEMPERATURE conditions will determine evaporation rates and hence water availability in the SOIL. The United Nations Environmental Programme (UNEP) has devised an aridity index, calculated by dividing the annual precipitation by annual potential evaporation. On this basis, 10% of the Earth's surface can be regarded as dry (aridity index 0.50–0.65), 18% is semi-arid (0.2–0.5), 12% is arid (0.05–0.20) and 8% is hyper-arid (<0.05).

Deserts are found mainly concentrated around the latitudes 30° north and south of the Equator, where the air masses that have been forced up by convection currents over the equatorial regions descend once more, having lost much of their water content in equatorial rains. [P.D.M.]

desertification The general reduction in the BIOMASS and PRODUCTIVITY of the world's drylands that has become increasingly apparent over the past few decades. [P.D.M.]

desertion 'Desertion' is usually used to describe abandonment of mating and parent–offspring associations, though the distinction between the two is not a clear one since desertion of a mate during offspring development amounts to desertion of the offspring as well. Despite its negative connotations, desertion may reflect adaptive life-history decisions on the part of the deserter. [C.J.B.]

desiccation The drying of living tissues, generally caused by exposure to air, and exacerbated if the air is dry and hot. [V.F.]

determinate growth Increase in body SIZE that ceases with maturation, such that there is no correlation between body size and age in adult individuals. [D.E.]

deterministic models, communities A deterministic model is a mathematical model of a process in which changes over time are determined by the initial state of the system and by the supplied values of driving variables, and do not depend on the outcome of chance events. The converse is a STOCHASTIC MODEL, in which some changes are attributed to chance. [M.O.H.]

deterministic models, populations Models in which a given input to the model always produces the same output. Models that predict the outcome of a process using PARAMETERS not subject to stochastic (random) variation. Deterministic models are generally mathematically more convenient than STOCHASTIC MODELS and their outcome is certain. Deterministic population models assume infinite POPULATION SIZE and ignore random FLUCTUATIONS in the environment with time. [V.F.]

detritivore *See* FEEDING TYPES, CLASSIFICATION.

detritophage *See* FEEDING TYPES, CLASSIFICATION.

detritus Non-living organic matter. Usually refers to particulate matter and, because it persists longer, to that of plant rather than animal origin, for example leaf litter. However, the term can also be applied to animal materials such as faeces and can sometimes be used to describe soluble organic materials. [P.C.]

detritus food chain DETRITUS is the primary source of energy and matter for some ecosystems, for example soils and sediments. Hence, detritivores are primary consumers in these systems. Often these detritivores rely on microbial communities, outside or within themselves, to render NUTRIENTS available from the hard-to-digest detritus. The detritivores may themselves be eaten by carnivores, that in turn could be eaten by secondary carnivores. So a detritus FOOD CHAIN might involve:

detritus → [microbe] → detritivore →
primary carnivore → secondary carnivore.

See also FEEDING TYPES, CLASSIFICATION. [P.C.]

deviance The measure of residual variation in statistical analyses that use maximum likelihood rather than least squares methods. Deviance is analogous to the residual SUM OF SQUARES (SSE) in least squares regression or ANALYSIS OF VARIANCE. [M.J.C.]

diadromous Migrating between saltwater and FRESH WATER. *See also* ANADROMOUS; CATADROMOUS. [P.O.]

diapause Diapause is a condition of arrested GROWTH or reproductive development common in many organisms, particularly insects, that live in seasonally varying environments. Diapausing stages may also show reduced metabolism and enhanced RESISTANCE to adverse climatic factors, such as cold, heat or DROUGHT. [J.S.B.]

diel Referring to events or actions that occur with a 24-h periodicity, for example migrations of planktonic animals, changes in oceanic photosynthetic potential and changes in near-shore PLANKTON communities in response to the tidal cycle. Often used synonymously with the term 'DIURNAL' (i.e. diurnal TIDES have one high tide and one low tide each day). [V.F.]

diet switching Several aspects of potential prey influence their inclusion in a predator's diet (*see* CONSTRAINTS ON FORAGING; DIET WIDTH). One obvious factor is their abundance. The change in the number of prey attacked by a PREDATOR within a given period as the density of the prey increases is known as the FUNCTIONAL RESPONSE. Functional response curves can take various forms depending on the constraints on a predator's intake rate. Where the constraints are SATIATION and HANDLING TIME (*see* DIET WIDTH), functional responses describe either a sharply truncated (type 1 response) or smoothly decelerating (type 2 response) curve with increasing prey density. In some cases, however, the contribution of a given prey type to the predator's diet also depends on the relative abundance of other types of prey. Many predators, particularly vertebrates, can learn to concentrate their SEARCHING effort on the most abundant prey. The result can be a sigmoidal (type 3) functional response, in which the proportion of a prey type in the diet is lower than expected from random sampling when its relative abundance is low but greater than expected when its abundance is high. The flip from underrepresentation to overrepresentation in the diet has been termed 'SWITCHING'. [C.J.B.]

diet width The range of items included in an organism's diet. [C.J.B.]

difference equations Difference equations are equations that involve finite differences. In ecology these are often used as population models for organisms having relatively DISCRETE GENERATIONS. In this application, the population at one time is determined by the population some fixed time previously. An example is the logistic map, which relates population at time $t+1$ (N_{t+1}) to population at time t (N_t):

$$N_{t+1} - N_t = rN_t(1 - N_t/K)$$

where r is the INTRINSIC RATE OF INCREASE of the population and K is its CARRYING CAPACITY. [S.N.W.]

differential equations Differential equations are equations involving not only variables but also some of the derivatives of those variables with respect to each other. A typical ecological example might be the ordinary differential equation:

$$\frac{dN}{dt} = f(N)$$

to describe the way in which the rate of change of population abundance with time, dN/dt, is related to abundance, N, through the function $f()$. Another example is the partial differential equation:

$$\frac{\partial N}{\partial t} = rN + D\left(\frac{\partial^2 N}{\partial x^2} + \frac{\partial^2 N}{\partial y^2}\right)$$

for a population growing exponentially, subject to diffusive movement. Stochastic differential equations involve additional stochastic terms, but their use is complicated by theoretical difficulties associated with taking limits of differences of stochastic processes. *See also* POPULATION DYNAMICS. [S.N.W.]

differential resource utilization The exploitation of different RESOURCES by the coexisting species. [P.O.]

diffuse competition Community-wide INTERSPECIFIC COMPETITION among a group of species, subjecting each species to a range of competitive pressures exerted by other species. The term is applied to trees in tropical forests, for example, where the species composition of neighbouring trees usually differs from one individual of a species to another. [J.G.S.]

dilution effect An advantage of living in a group, whereby a member of a group dilutes the impact of an attack by a PREDATOR that can kill only one group member per successful attack. *See also* ALLEE EFFECT. [P.O.]

dioecy A species is dioecious if its population contains both male and female individuals and no HERMAPHRODITES (in the broad sense). [I.O.]

dioxin Generic term for a family of chlorinated hydrocarbons including polychlorinated dibenzo-*p*-dioxins and furans. They are formed primarily during combustion of chlorinated organic materials (e.g. chlorinated solvents, plastics), although vehicle emissions and coal burning are also sources. The different congeners have very different toxicities, with some being among the most toxic substances known, so that concentrations as low as 10^{-12} g m^{-3} in air, or less, need to be measured for effective monitoring of human health risks. [J.N.C.]

diploidy A descriptor of the diploid state (two copies of homologous genome) that primarily refers to the cell nucleus. The cell nucleus is haploid when it includes a single copy of the genome. Haploid and diploid nuclear phases alternate during the LIFE CYCLE of all sexualized EUKARYOTES. Haploidy and diploidy do not refer only to the ploidy level of the cell nucleus but also to that of a given organism or a group of individuals. For example, mammals are diploid because almost all their cells (except the GAMETES) are diploid. [M.V.]

diplontic life cycle *See* SEXUAL REPRODUCTIVE CYCLES.

direct fitness In KIN SELECTION theory, that part of an individual's inclusive FITNESS gained through its direct investment in its own offspring (as opposed to the offspring of relatives it may have helped to raise). The direct fitness of an individual does not include the portion of its own offspring that it gained through the actions of others. [T.J.K.]

direct transmission, parasites The process by which a parasite is passed from one (infectious) HOST to another (susceptible) host without any intervening mechanism. [G.F.M.]

directed selection *See* SYMPATRIC SPECIATION.

directional selection NATURAL or ARTIFICIAL SELECTION favouring phenotypes at one extreme of the distribution. As a result of directional selection, the population MEAN of the selected TRAIT is expected to shift in the favoured direction. [T.J.K.]

disassortative mating Mating with phenotypically dissimilar individuals. *See also* MATING SYSTEMS. [P.O.]

disclimax In the scheme of Clements (*see* CHARACTERS IN ECOLOGY) a kind of VEGETATION differing from the climatically determined CLIMAX as a result of the effects of humans and/or their domestic animals. [P.J.G.]

discrete generations POPULATIONS of a single species in which generations do not overlap and are therefore distinguishable from each other. [A.J.D.]

discrete size-classified population model *See* SIZE DISTRIBUTIONS.

disharmony A characteristic of the fauna and flora of isolated islands, which often completely lack elements that are conspicuously present on continents, for example the only mammals endemic to the Hawaiian islands were one species of bat and one species of seal. *See also* ISLAND BIOGEOGRAPHY. [P.O.]

disjunct distribution A given TAXON (species, genus, family, etc.) is said to possess a disjunct distribution when its geographical RANGE comprises two or more populations that are separated by distributional gaps greater than the normal DISPERSAL capacity of the taxon. [J.H.T.]

disjunct populations Populations that are geographically isolated from the rest of the TAXON. There are two types of disjunction: (i) undifferentiated disjunction refers to disjunct populations that are similar; and (ii) differentiated disjunction occurs when one or more of the disjunct populations diverge from the others. Disjunction is of considerable evolutionary interest because it is believed to facilitate SPECIATION. [V.F.]

dispersal Dispersal is the movement of individuals away from where they were produced and may be active or passive. *Cf.* DISPERSION. *See also* ACTIVE DISPERSAL; PASSIVE DISPERSAL. [R.H.S.]

dispersal polymorphism DISPERSAL may result from the day-to-day activities of an organism or be a specific activity that leads the organism to move outside of its typical foraging range. In the latter case, it has been observed that the tendency to disperse is not randomly distributed among individuals within the population but is inherited to a greater or lesser degree. In the simplest case dispersal might be controlled by a single locus but, more generally, several to many genes are likely to be involved. [D.A.R.]

dispersion Dispersion is a statistical term meaning variation or spread and is not to be confused with DISPERSAL, which is a biological process. In ecology, dispersion usually refers to the spatial DISTRIBUTION of organisms. Unfortunately, use of words based on dispersion has led to some confusion in the literature. The term 'overdispersed', for example, may imply one thing to a statistician (a distribution with a high variance, as when organisms occur in clusters) and the opposite to a behavioural ecologist (dispersal to such an extent that organisms are very spaced out); 'underdispersed' causes similar problems. For this reason, use of the terms 'overdispersed' and 'underdispersed' is best avoided.

Three main types of dispersion pattern are recognized. When individuals are very evenly spread, perhaps because any individual actively avoids being close to other individuals (e.g. red grouse defending territories), their distribution is said to be uniform or regular (*see* REGULAR DISTRIBUTION). When individuals occur in clumps or clusters, perhaps because they are attracted to a patch of food (e.g. carrion flies on a carcass), their distribution is said to be contagious or aggregated (*see* AGGREGATED DISTRIBUTION). A so-called RANDOM DISTRIBUTION occurs if there is an equal chance of any individual occupying any point in space and individuals are distributed independently of one another. Analysis of dispersion patterns is conventionally based on either distances between individuals, counts in randomly placed quadrats or 'units of habitat', such as fungal bodies, fruit or carcasses. [R.H.S.]

displays Communicatory acts between individuals of the same species; often take the form of elaborate, and/or bizarre, activities. Male fiddler crabs (*Uca* spp.) wave a grossly enlarged claw in the air to attract females, male ducks court by performing sham drinking and preening actions, and lizards of the genus *Scleroporus* bob their heads vigorously to threaten rivals. Such displays can be puzzling in their complexity and in the seeming irrelevance of some of their elements to the task in hand. However, much can be understood about displays by considering their motivational ancestry. [C.J.B.]

disposable soma theory A theory that explains AGEING (SENESCENCE) through asking how best an organism should allocate its metabolic RESOURCES (primarily energy) between, on the one hand, keeping itself going from one day to the next (maintenance) and, on the other hand, producing progeny to secure the continuance of its genes when it itself has died (reproduction).

No species is immune to hazards of the environment, such as predation, starvation and disease. These hazards limit the average survival time, even if senescence does not occur. It follows that maintenance is only needed to an extent which ensures that the body (soma) remains in sound condition until an age when most individuals will have died from accidental causes. In fact, a greater investment in maintenance is a disadvantage because it eats into resources that in terms of NATURAL SELECTION are better used for reproduction. The theory concludes that the optimum course is to invest fewer resources in the maintenance of somatic tissues than would be necessary for indefinite somatic survival. [T.B.L.K.]

disruptive selection NATURAL or ARTIFICIAL SELECTION favouring phenotypes that deviate in either direction from the population MEAN. [T.J.K.]

dissolved organic matter (DOM) Dissolved molecules derived from degradation of dead organisms or excretion of molecules synthesized by organisms. In practice, DOM is defined arbitrarily to include all organic matter passing through a 0.45-μm filter. [V.F. & P.C.]

dissolved oxygen (DO) Oxygen in aquatic environments that is freely available to support RESPIRATION. [P.C.]

dissolved solids *See* DISSOLVED ORGANIC MATTER.

distribution Within ECOLOGY this term in used in two slightly different contexts. It can be used to describe the spatial distribution of the individuals in a POPULATION (*see* AGGREGATED DISTRIBUTION; RANDOM DISTRIBUTION; REGULAR DISTRIBUTION). It can also be used to describe the distribution of any set of measurements or observations (*see* PROBABILITY DISTRIBUTIONS). In some situations, a particular probability distribution can be used to describe the spatial distribution of individuals (*see* NEGATIVE BINOMIAL DISTRIBUTION; POISSON DIS-

TRIBUTION). In this case the probability distribution attempts to describe the spatial distribution. *See also* DISPERSION. [B.S.]

disturbance

1 Any process that destroys plant or animal BIOMASS (*sensu* Grime); effects of animals (GRAZING, trampling and burrowing), fire and bulldozer are all agencies of disturbance. One of the two important axes of GRIME'S TRIANGLE, the other being environmental FAVOURABILITY.

2 The creation of a seedbed by physical removal of vegetative cover, possibly accompanied by tillage of the soil (e.g. break-up of the soil crust in arid ecosystems). Natural agencies of disturbance include landslide, hurricane, fire, flood, silt deposition, glaciation, etc. Digging animals (rabbits), tunnelling species (moles, earthworms) and colony-dwelling species (gophers) are important biotic agents of disturbance, creating open conditions for seedling ESTABLISHMENT. In many ecosystems, the principal agents of disturbance are humans and their machines (agricultural cultivation, forest clearance, construction works, etc.).

3 The creation of microsites (safe sites) for RECRUITMENT of plant species, by whatever means (reduction of perennial plant cover, provision of bare soil or canopy gaps suitable for seedling establishment).

4 The creation of gaps in a forest canopy by removal or death of the dominant trees where sapling recruitment can occur (gap phase regeneration).

Disturbance creates space in previously CLOSED COMMUNITIES and provides opportunities for establishment; thus disturbed communities are particularly susceptible to invasion by new species. [M.J.C. & K.T.]

diurnal Most terrestrial organisms are exposed to daily cycles of light and dark to which they become entrained, i.e. adapted. Many physiological (e.g. flowering) and behavioural (e.g. drosophilid eclosion) patterns are determined by specific periodic responses to such external cues. When a particular activity or response occurs during the light phase then it is described as diurnal (or circadian). Alternatively it may occur at dusk (crepuscular) or during the dark phase (nocturnal). [R.C.]

diversity The terms 'diversity', 'ecological diversity', 'species diversity', 'biological diversity' and 'BIODIVERSITY' all refer to the variety and ABUNDANCE of species at a specified place and time. Diversity is a concept that is intuitively easy to understand but remarkably difficult to quantify (*see* DIVERSITY INDICES). The reason for this is that diversity consists of not one but two components: SPECIES RICHNESS, i.e. number of species, and EQUITABILITY (sometimes termed 'evenness'), which is a measure of how equally abundant those species are. Communities with a large number of species are obviously more diverse than species-poor ones. However, high equitability, which occurs when species are equal or virtually equal in abundance, is also equated with high diversity. For example, a sample of 100 moths with 10 individuals in each of 10 species is considered more diverse than another which also has 100 moths and 10 species, but in which one of the species is represented by 91 individuals. These two components of diversity may be evaluated either separately (*see* EQUITABILITY; SPECIES RICHNESS) or jointly by means of a composite measure such as the Shannon or Simpson index. [A.E.M.]

diversity, alpha, beta and gamma Three measures of species DIVERSITY in space.

All species have a distinctive ecological specialty that allows them to co-occur with others. Some of these specializations are founded on temporal differences in the environment, some on spatial differences, some on having different victims or enemies or mutualists, and some on differences in the ability to tolerate poorer conditions or dominate better ones.

The number of species at a single point in space and time is alpha diversity. It estimates how many species co-occur because they specialize on different interacting species. For example, differences in body size between species often lead to differences in which species consume them or which species they eat; so species of various body sizes often co-occur at a point and contribute to alpha diversity. Temporal differences can also contribute to alpha diversity because species may decline very slowly after their special season or year has passed.

As we sample areas larger than a point, additional habitats are included and diversity grows. The rate of growth is beta diversity. We have no standard formula to calculate beta diversity, but the most common uses the coefficients of the SPECIES–AREA RELATIONSHIP: $S = cA^z$, where S is the number of species and A is the area (*see* DIVERSITY GRADIENT).

The gamma diversity is the value of S in an area. It combines point diversity with the effect of having many habitats in space. [M.L.R.]

diversity gradient A CORRELATION of DIVERSITY with another spatial or temporal variable. [M.L.R.]

diversity indices There are three major groups of diversity index: (i) species-richness measures (count number of species present); (ii) SPECIES-ABUNDANCE MODELS with an associated diversity index; and (iii) indices based on proportional abundance of species that incorporate both properties of diversity (richness and relative abundance). The choice of appropriate index from the bewildering variety available depends on such factors as difficulty in appraisal of species abundance and the success in sampling and identifying all species present. [P.S.G.]

division of labour Different units/organisms (modules) in a colonial, modular or SOCIAL ORGANIZATION have different biological/ecological roles (e.g.

in terms of FEEDING, defence and reproduction). *See also* MODULAR ORGANISM. [P.C.]

Dollo's law The proposition that EVOLUTION is irreversible or that past states of an evolutionary LINEAGE can only be at most partially re-created. Named after L. Dollo (1893). A variant on this proposition is that an organ, once lost, will never again be present in its original form. [L.M.C.]

DOM *See* DISSOLVED ORGANIC MATTER.

domestication Historical and evolutionary changes in plants and animals when brought under human household (Latin *domus*, home) care and uses. [S.K.J.]

dominance

1 In BEHAVIOURAL ECOLOGY the term is used to refer to the social status of an animal within a group. An individual that maintains high social status by aggressive behaviour towards others is exhibiting dominance. *See also* DOMINANCE, SOCIAL.

2 In VEGETATION studies, dominance refers to the capacity for one species to exert overriding influence upon others within the COMMUNITY. Usually the dominant plant species is the one that has the highest BIOMASS in the ECOSYSTEM and which therefore controls the microclimatic conditions for all others. Occasionally, however, dominance may be exerted by a plant of lower stature, which may control potential invaders or the regeneration of more robust species by competitive means. Therefore, it may be difficult to determine dominance by visual inspection and the use of the term then becomes speculative.

3 In genetics, refers to ALLELES whose effects are expressed in the PHENOTYPE.

[P.D.M.]

dominance-controlled communities Communities where competitive interactions between species form a strong hierarchy in which inferior competitors are displaced by the arrival of superior competitors resulting, in the absence of DISTURBANCE, in the development of the COMMUNITY to a predictable end-state dominated by the competitively superior species. [P.H.W.]

dominance hierarchy Dominance hierarchies are group-level constructs that reflect competitive relationships between individuals within social AGGREGATIONS (where aggregations may be temporary gatherings or cohesive social groups). They arise where competitive interactions between a given pair of individuals are resolved, sometimes quickly and with little aggression, with a consistent bias in favour of one individual. Individuals can thus be ranked into a hierarchy according to their competitive priority (DOMINANCE) over other members of the aggregation. [C.J.B.]

dominance, social Social DOMINANCE may best be defined as a stable asymmetry or predictability in the outcome of aggressive interactions between two or more individuals. [D.A.D.]

donor-controlled systems These are 'ideal' systems where the POPULATION SIZES of species at a particular TROPHIC LEVEL are entirely controlled by the ABUNDANCE of species in the trophic level below it. In a simple example of a donor-controlled system, the density of grasses controls the density of herbivores which, in turn, controls the density of predators. However, herbivore density is independent of predator density and grass density independent of herbivores. While such systems are not entirely realistic there are some that appear to follow this pattern quite well and they do make population modelling considerably easier. *See also* POPULATION DYNAMICS; POPULATION REGULATION. [C.D.]

dormancy Resting condition with relatively reduced metabolism that might involve the whole organism, as in higher plants and animals, or be confined to propagules, such as resting spores in fungi and bacteria, resting EGGS in some animals, non-germinating seeds and non-growing buds in plants. Also referred to as hypobiosis. *See also* AESTIVATION; DIAPAUSE. [P.C.]

dose–response The quantitative relationship between the amount of a TOXICANT administered or taken (usually by feeding or injection) or absorbed by a subject and the incidence or extent of adverse effects. [P.C.]

drift, evolutionary Random FLUCTUATIONS of gene frequencies in populations caused by random sampling of GAMETES and/or GENOTYPES in reproduction. Drift is most important in small populations, but the fate of rare ALLELES is determined by random sampling also in very large populations. Drift induces fixation and loss of alleles and, thereby, can cause a decay of genetic VARIABILITY. [W.G.]

drift, freshwater When benthic organisms of streams or RIVERS enter the water column and are moved downstream, the phenomenon is known as drift. [R.W.D.]

drought Drought has no universally accepted definition, although it is normally recognized as a sustained and regionally extensive state of water deficiency. It is frequently associated with famine and aridity. [M.B.]

dry mass The mass of a part of an organism, a whole organism or all the organisms from a given area after removal of its moisture by evaporation. [S.P.L.]

dry season A period of the year when precipitation is greatly reduced or absent. The term is particularly used of tropical and SUBTROPICAL regions in which precipitation (often monsoonal) is confined to particular times of the year and the remainder of the year is dry. [P.D.M.]

d_x In a LIFE TABLE the proportion of individuals that die in each AGE CLASS or in each stage. Calculated as $l_x - l_{x+1}$. The values of d_x are additive so that the total MORTALITY at ages 3, 4 and 5 is $d_3 + d_4 + d_5$ but, unlike q_x and k_x, they give little indication of the intensity of mortality at each age since, even with the same mortality rate, *d* will be large if there is a large number of organisms present and at risk of dying. [A.J.D.]

dynamic equilibrium A dynamic equilibrium occurs when some characteristic of a system is in an unchanging state, despite the fact that its constituent parts may be changing. For example, the size of a population in which IMMIGRATION and birth exactly balances EMIGRATION and death will be in dynamic equilibrium. The individuals making up the population are changing constantly and the birth, death and migration rates may all be varying, but the size of the population is constant. [S.N.W.]

dynamic programming A method of modelling dynamic decision-making problems in which current choices are likely to influence options in the future. The approach relies on knowing the required end-state at the outset, then working back to the starting state seeking appropriate choices at each decision point. [C.J.B.]

dynamically fragile If the dynamics that a system displays are very sensitive to parameter changes or small perturbations then the system is dynamically fragile. *Cf.* DYNAMICALLY ROBUST. [S.N.W.]

dynamically robust Dynamics that are fairly insensitive to parameter changes or perturbations are robust: the same patterns will occur for a relatively wide range of circumstances. *Cf.* DYNAMICALLY FRAGILE. [S.N.W.]

dystrophic A term applied to a shallow freshwater LAKE in which the presence of organic materials lends a dark brown colour to the water. Such lakes are of low biological PRODUCTIVITY and have poor light penetration. Cf. eutrophic and oligotrophic (*see* EUTROPHICATION). [P.D.M.]

E

EC_{50} Statistically derived concentration of XENOBIOTIC that has a defined adverse effect (often behavioural, e.g. 'dancing' movement in *Daphnia*) in 50% of an observed POPULATION over a prescribed time in defined conditions. Also referred to as median effective concentration. Can also refer to exposures in terms of dose (ED_{50}). [P.C.]

ecological constraint hypothesis The hypothesis that certain individuals aid in the reproductive efforts of others, either as 'helpers at the nest' in cooperative breeders or as 'workers' in eusocial species, because they are restricted in their ability to disperse or breed on their own and are thus 'making the best of a bad job'. The alternative is that helpers or workers experience higher FITNESS by aiding the breeding efforts of others compared to breeding independently because of intrinsic advantages derived via helping. [W.D.K.]

ecological efficiency Ecological EFFICIENCY (also called LINDEMAN EFFICIENCY or transfer efficiency) is a measure of the proportion of the energy assimilated by one TROPHIC LEVEL that is then assimilated by the trophic level above it. This is thus a measure of the efficiency with which energy is transferred between levels of a trophic pyramid and is defined as:

$$\text{Ecological efficiency} = \frac{\text{Assimilation at trophic level}_n}{\text{Assimilation at trophic level}_{n-1}}$$

It was originally thought that this might be a predictable constant for different trophic levels in different ECOSYSTEMS and it was suggested that approximately 10% was a widespread value, based on laboratory studies of water fleas (*Daphnia* spp.) feeding on phytoplankton. Further research has shown that it is neither constant nor clearly predictable; values can vary from *c.* 1%, for example for wolves feeding on moose on Isle Royale in Lake Superior up to 70% in some marine FOOD CHAINS.

One consequence of low ecological efficiencies is that organisms at the base of the FOOD WEB are much more abundant than those at higher trophic levels, which explains the shape of pyramids of numbers and biomass. These illustrate graphically the consequences of the rapid loss of energy between plants and herbivores and between herbivores and carnivores, which is a biological manifestation of the second law of thermodynamics.

The ecological 'inefficiency' is partly due to energy being lost from the system as heat, usually monitored by measuring the RESPIRATION of members of a trophic level, and partly because no trophic level is completely assimilated by the level above it: some energy is always diverted into the decomposer pathways either as dead bodies and/or as undigested faecal remains. The realization that a majority of the energy incorporated as net PRODUCTIVITY by members of one trophic level usually flows into these saprophytic pathways has promoted much research into the ecology of decomposers. *See also* ASSIMILATION EFFICIENCY; ECOLOGICAL PYRAMID; PRODUCTION EFFICIENCY. [M.H.]

ecological energetics Ecological energetics concerns the transfer of energy that occurs within ECOSYSTEMS measured at the level of the individual organism, populations, group of populations or TROPHIC LEVEL. Detailed studies of the transformations of energy that occur within the cells of an organism are usually considered to be the realm of physiological energetics. The ecologist interested in energetics is primarily concerned with the quantity of incident energy per unit area of the ecosystem and the efficiencies with which this energy is converted by organisms into other forms. The term BIOENERGETICS has been taken by some to subsume ecological energetics, but by others it has been used more restrictively to refer to the energetics of domestic (especially agricultural) animals, and by others to refer to cell energetics. [M.H.]

ecological experiments Ecological, or field, experiments bridge the gap between laboratory experiments and observations, and simple observations and measurements of the natural world. The attraction of ecological experiments is realism: the full complexity of the environment in which animals and plants live remains, yet variables of interest are manipulated in a controlled and replicated manner so that hypotheses about their role can be tested. One of the most common types of ecological experiment is where the densities of a particular species are manipulated in some way and the responses of the COMMUNITY are observed. DENSITY manipulations are achieved in a variety of ways including, for example, fishing, removal or addition by hand or spraying with insecticides. An important objective for the experimenter is to minimize and control for the effects of the method itself. For plants and sessile or slow-moving animal species, or when the HABITAT itself has a natural boundary (e.g. islands or lakes), there is little difficulty in maintaining desired densities. In other cases, however, it is often necessary to use physical or chemical barriers, such as fences, cages or toxic paint, to surround experimental plots and keep species at higher and/or lower densities than those in the unmanipulated environment. Designing adequate controls for the effects of such structures is a key part of field experimentation. Other types of treatments in ecological experiments include the MANIPULATION of nutrient or light levels, or the degree of DISTURBANCE. [S.J.H.]

ecological genetics Ecological geneticists combine field and laboratory studies to investigate the adjustments and ADAPTATIONS of wild populations to their environment. Their work has concentrated on the EVOLUTION of a number of adaptive TRAITS, for example industrial MELANISM in moths, shell colour and banding variation in *Cepaea* snails, insecticide RESISTANCE in insects, heavy-metal TOLERANCE in plants, BATESIAN MIMICRY in butterflies, wing spotting in meadow brown butterflies and chromosome inversions in fruit flies and mice. [P.M.B.]

ecological indicator Organism(s) whose presence indicates occurrence of particular set of conditions. *See also* BIOTIC INDICES. [P.C.]

ecological isolation

1 Two broadly coexisting species are said to be ecologically isolated when their normal environmental requirements are significantly different. In ecology, this usage often occurs in discussions of INTERSPECIFIC COMPETITION for resources. In such cases, ecological isolation is often invoked to account for the observed COEXISTENCE of closely related species.

2 Sympatric organisms of two species are said to be ecologically isolated when they are constrained from meeting, and hence conceivably mating, by their different environmental requirements. Ecological isolation is one of the pre-mating ISOLATING MECHANISMS recognized by the authors of the BIOLOGICAL SPECIES CONCEPT. Darwin was aware of ecological isolation in terms of REPRODUCTIVE ISOLATION. Ecological isolation in this sense is thought by some to be an incidental consequence of geographically DISJUNCT POPULATIONS adapting to distinct habitats to which they have been contingently restricted. Thus, ecological isolation does not evolve as an ADAPTATION with the function of preventing hybridization as the term 'isolating mechanism' might suggest. [H.E.H.P.]

ecological pyramid Ecological pyramids are ways of conceptualizing the structure of natural communities of interacting species. They are all based on the fundamental premise that members of a COMMUNITY can be compartmentalized into trophic levels by classifying them according to how they derive their energy, in the case of animals by what they eat.

If the organisms in each TROPHIC LEVEL are scored in some way, for example by simply counting them (PYRAMID OF NUMBERS), counting and weighing them (pyramid of biomass) or measuring the rate of energy transfer through them (pyramid of energy), with the scores for each trophic level displayed along a horizontal axis and the levels stacked vertically with the primary producers as the base, it is usually found that the resulting diagram represents a pyramid. Several generalizations stem from this observation. One is that prey items are usually more numerous or have a greater BIOMASS per unit area than the consumers that eat them. Another is that this illustrates the second law of thermodynamics by showing that the EFFICIENCY of energy transfer through trophic levels is never perfect but that each transformation is accompanied by a significant loss of energy as heat. *See also* ECOLOGICAL EFFICIENCY; ECOLOGICAL ENERGETICS. [M.H.]

ecological restoration Ecological restoration is the return of an ECOSYSTEM to an approximation of its structural and functional condition before damage occurred. [J.C.]

ecological rucksack The total mass of material flow 'carried by' an item of economic consumption (product) in the course of its life cycle. [P.C.]

ecological species concept A lineage (or closely related set of lineages) which occupies an adaptive zone minimally different from that of any other lineage in its range and which evolves separately from all lineages outside its range'. *Cf.* BIOLOGICAL SPECIES CONCEPT. [L.M.C.]

ecologism Use of ecological terms and concepts (often superficially and/or naively) in political/moral debate. Also refers to the terms that are used in that way. [P.C.]

ecology Ernst Heinrich Haeckel is usually attributed with originating the term from the Greek *oikos*, house or dwelling place (*see* CHARACTERS IN ECOLOGY).

It began as a subject concerned predominantly with the ENVIRONMENT (Haeckel's 'house'), i.e. the study of the way that organisms are influenced by physicochemical conditions. However, as it has developed, it has become increasingly concerned with interactions between individuals, so that it is now probably best defined as that area of biology concerned with the study of collective groups of organisms. As such, it stands at the opposite end of the biological scheme of things from the study of cells and molecules.

The spheres of interest of ecology and ecologists therefore range from individual organisms, through populations, to communities and ECOSYSTEMS. Subdisciplines emphasize interests in different elements of these interactions: PHYSIOLOGICAL ECOLOGY looks at the way physiology influences

DISTRIBUTION and ABUNDANCE and is influenced by environment; BEHAVIOURAL ECOLOGY is concerned with the way behaviour responds to ecological challenges in both the short term and the longer (evolutionary) term; POPULATION ECOLOGY describes and explains distribution and abundance; COMMUNITY ECOLOGY describes and explains species associations; ecosystem ecology describes and explains fluxes of energy and matter as with communities. The study of individuals to populations is sometimes referred to as AUTECOLOGY, whereas the study of communities and ecosystems is sometimes referred to as synecology. There are numerous other descriptions: functional ecology is physiological ecology but emphasizes an experimental and hypothesis-testing approach; trophic ecology studies feeding ecology; MOLECULAR ECOLOGY uses techniques from molecular biology to address ecological questions; ECOLOGICAL GENETICS tries to understand gene frequencies in terms of ecological processes, etc. And, of course, it can also be subdivided with respect to HABITAT (marine, fresh water, soil, littoral, benthic, etc.) and organisms (microbial, animal, plant, avian, fish, etc.).

Applied ecology involves using understanding from all these various aspects of ecology in the protection, conservation and management of ecological systems. Theoretical ecology usually involves the development of broad understanding with emphasis on generality, whereas applied ecology is often concerned with the protection or exploitation of particular systems and with attempts to make precise quantitative predictions about them; linking these two endeavours is not always easy. [P.C.]

ecomorphology The ecomorphological hypothesis states that morphometric data serve as an index of ecological characteristics such as NICHE WIDTH. [S.J.H.]

ecophysiology Use of physiological observations on organisms to help explain their DISTRIBUTION and ABUNDANCE. *See also* PHYSIOLOGICAL ECOLOGY. [P.C.]

ecosystem A functional ecological unit in which the biological, physical and chemical components of the ENVIRONMENT interact. This term focuses attention on the complex interplay between plants and animals and ABIOTIC FACTORS of their HABITAT. It is a much wider concept than the COMMUNITY with which it is sometimes confused. [A.J.W.]

ecosystem engineers Organisms that directly or indirectly modulate the availability of RESOURCES (not themselves) to other species, by causing physical changes in the biotic or abiotic materials of the environment. In so doing they create or modify HABITATS. ALLOGENIC engineers change the environment by transforming living or non-living materials from one physical condition to another via mechanical or other means, for example beavers and earthworms. AUTOGENIC engineers alter environments by their own physical structure, such as the growth of a forest or coral reef.

The term 'engineer' needs care. Clearly it is intended as an analogy; however, 'engineers' work to 'design' to produce artefacts for the benefit of others. The dictates of NATURAL SELECTION are such that the ECOSYSTEM 'engineering' has to bring FITNESS benefits to the 'engineers' and not for the good of the group (*cf.* GROUP SELECTION). The analogy needs to be explicit that engineers are usually compensated by fees for their work. But even this is somewhat forced, since the fees will usually be related to the quality and quantity of the work, whereas the fitness returns to the ecosystem 'engineers' will not necessarily be proportional to the benefits gained by the 'beneficiary' species. [P.C.]

ecosystem health It is relatively easy to know when people are ill; they develop disease-specific symptoms, feel off-colour and function abnormally. We can gauge the seriousness of this in terms of various general indicators such as body temperature and pulse rate, the extent to which normal function is impaired and also in terms of the likelihood of recovery. Increasingly, some scientists and policy-makers are describing the condition of ECOSYSTEMS in terms of states of health. Yet there are some important differences between human bodies and ecosystems. The latter are not as obviously and tightly organized as organisms. Where, for example, are the coordinating nerve centres of ecosystems? Individuals and species can be lost from ecosystems without apparently impairing normal function to an extent that cells, tissues and organs could not be lost from organisms. So ecological norms are not so easily defined as body norms; it is more difficult to identify the 'pulses' of ecosystems. Also it seems that their patterns of recovery after DISTURBANCE may not always be very predictable and the recovered systems may never be quite the same as the originals. [P.C.]

ecosystem integrity The view that a number of key states (properties and processes) must be intact for an ECOSYSTEM to persist in a stable state. [P.C.]

ecosystem management The science and art of directing human activities to sustain or restore the desired DIVERSITY and PRODUCTIVITY of terrestrial and aquatic ecosystems in an area. [A.N.G. & T.A.S.]

ecosystem redundancy Removal of some species does not necessarily lead to the breakdown of function in an ECOSYSTEM. Hence it is presumed that their role is redundant (i.e. unneeded functional capacity) in 'normal' ecosystems. [P.C.]

ecosystem services Benefits that society obtains from properly functioning ECOSYSTEMS in terms of the supply of raw materials, food, clean water, clean atmosphere, etc. [P.C.]

ecotone The zone of transition between (different) adjacent ecological systems (ECOSYSTEMS, COMMUNITIES, HABITATS) with characteristics uniquely defined by time and space and by the strength of the interaction between the adjacent ecological systems. [B.W.F. & K.T.]

ecotope Term used in an evolutionary context to describe the full range of NICHE and HABITAT factors that affect a species and determine its survival. To survive, a species must have a unique ecotope, either with respect to niche or habitat factors or both. [B.W.F.]

ecotoxicology A subject area concerned with understanding where anthropogenic chemicals go in the environment (their fate) and hence the extent to which ecological systems are exposed to them (exposure) and, in consequence, the ecological effects that they have (effects). It can be carried out retrospectively (are releases having effects?) and prospectively (are releases likely to have effects?). Retrospective analyses involve MONITORING programmes; prospective studies involve predictive tests. Ecotoxicology is therefore a multidisciplinary/interdisciplinary subject involving a combination of environmental chemistry, toxicology and ecology. It is being used increasingly as a basis for environmental RISK ASSESSMENT in the development of environmental protection legislation. [P.C.]

ectoparasite A PARASITE that lives on the outside of the HOST (used only in reference to macroparasites; *cf.* ENDOPARASITE). [G.F.M.]

ectotherm An ectotherm (also called POIKILOTHERM or allotherm) is an organism that relies on the TEMPERATURE of its surrounding environment to regulate and maintain its own internal body temperature. [R.C.]

ED_{50} *See* EC_{50}.

edaphic Edaphic factors are the physical, chemical and biotic characteristics of the SOIL that influence plant growth and distribution. [J.R.E.]

edge effect Edges are structural or compositional discontinuities between two adjacent patch types. Among the most familiar form of edges in temperate zones are boundaries between forests and clearcut areas or agricultural fields. However, edges can also be of natural origin. Edge environments occur naturally at many ECOSYSTEM boundaries: along the perimeter of bodies of water, cliffs, outcrops of exposed rock, and sharp discontinuities in soil type or HYDROLOGY. [D.A.F.]

effective ecosystem size The 'effective size' of an ECOSYSTEM depends on its absolute size (extent) and on the degree of fine-grained physical structure within it. The term refers to the characteristics of an ecosystem that cause it to be functionally larger (for the organisms in it) than it actually is. Ecosystems of greater 'effective size' reduce the risk of predator or prey extinctions by limiting the amplitude of density OSCILLATIONS. [S.J.H.]

effective population size The effective population size of an actual population can be seen as a measure of how many genetically distinct individuals actually participate in the formation of the next generation and as a predictor of the rate at which, because of drift, genetic VARIABILITY will be lost from a population or inbreeding will increase. It has thus applications in CONSERVATION BIOLOGY, as it might be used to produce the minimum population size under which a given species might lose its evolutionary potential. [I.O.]

efficiency Ratio of a useful output to an input. Used widely to describe ecological systems at various levels. [P.C.]

effluents Waste release from point sources, usually as liquid into external environment. [P.C.]

egg mimicry The close resemblance of the appearance of the EGGS of some brood-parasitic birds to that of their HOSTS. [A.P.M.]

eggs Unicellular products of female reproductive system. Usually formed by meiosis and often requiring conjugation with male contribution to induce development. However, the latter need not be the case in PARTHENOGENESIS. The cytoplasm often contains some material (yolk) that supports early development. [P.C.]

El Niño Episodic climatic changes that include warming of the equatorial Pacific Ocean and suppression of UPWELLING into the EUPHOTIC ZONE off the coast of Peru by intrusions of warm, nutrient-poor, surface water. The southward-flowing tongue of the equatorial countercurrent is a regular phenomenon in northern Peru and occurs every year in February or March. Periodically the current extends further south to >12°S and displaces the cold northward-flowing Peru current. El Niño is associated with a weakening of the trade winds, which occurs typically in cycles of 7 years and may be caused by global circulation anomalies that are in turn influenced by SOLAR RADIATION. The El Niño has been responsible for mass mortality of PLANKTON and fish, has led to starvation of seabirds and has contributed to the collapse of the Peruvian anchovy fishery.

El Niño (derivation: Christ boy-child, as it typically occurs around Christmas) was named by Spanish-speaking fishermen off Ecuador and Peru who correlated its occurrence (warm surface-water temperatures) with disastrous fishing conditions. The Australian Bureau of Meteorology (1988) definition is: 'the occasional warming of the usually cool surface waters of the eastern equatorial Pacific'. Over the past 40 years, nine El Niños have affected the South American continent.

The concept has gradually been extended to cover the whole of the Southern Ocean and now beyond. [J.M. & V.F.]

Eltonian pyramid The idea of expressing the numbers in, and BIOMASS or mass of, different trophic levels as horizontal bars stacked in a pyramid is usually attributed to Charles Elton (*see* CHARACTERS IN ECOLOGY), who first drew such diagrams following his formative expedition to Spitsbergen. *See also* ECOLOGICAL PYRAMID. [M.H.]

emergence marsh The upper part of a SALT MARSH, which is subject to only short periods of submersion (immersion) during spring TIDES and where the key environmental conditions are associated with long periods of emersion. [B.W.F.]

emergent plant Aquatic plant species, usually growing in the shallower marginal regions of a freshwater body, which are rooted below the water surface but produce shoots that extend above the surface. [P.D.M.]

emergent properties Features not fully (or at all) explicable in terms of the properties of the component parts. Thus the dynamics of collective groups of organisms involve interactions between them that are not fully explicable on the basis of understanding of individuals and their parts (cells and molecules). [P.C.]

emigration Exit of conspecifics from a POPULATION to elsewhere. Emigration leads to GENE FLOW out of the population. *See also* DISPERSAL; IMMIGRATION. [P.C.]

endangered species Official designation is species that have 20% probability of becoming extinct in 20 years or 10 generations (US Endangered Species Act). *See also* MINIMUM VIABLE POPULATION. [P.C.]

endemic species An animal or plant that is native to a particular location and is restricted to that location in its DISTRIBUTION. [P.D.M.]

endemism *See* ENDEMIC SPECIES.

endobiont An organism such as a bacterium or alga that lives inside the cells of other organisms, often potentially an endosymbiont. [S.G.C.]

endogenous rhythm Internally generated rhythm (oscillation) that is self-sustained. A CIRCADIAN RHYTHM is an endogenous rhythm with a daily periodicity. [J.S.B.]

endolithic Organisms living with rocks. [A.P.]

endoparasite A PARASITE that lives within the HOST (used only in reference to macroparasites; *cf.* ECTOPARASITE). [G.F.M.]

endotherm An endotherm is an organism that is able to maintain and regulate its internal body TEMPERATURE across a wide range of conditions present in the surrounding external environment. [R.C.]

energy budget An energy budget is a statement of the balance between energy input and energy output at any point of energy transfer through an ECOSYSTEM, TROPHIC LEVEL, POPULATION or individual organism. It is usually formulated in the form of an equation, with the units usually used to form a common currency for different elements of the equation being joules or kilojoules. [M.H.]

energy–diversity relationship Correlations have been found on a global or regional scale between SPECIES RICHNESS and indices of availability of energy to organisms such as PRIMARY PRODUCTIVITY, evapotranspiration and availability of food or limiting NUTRIENTS (e.g. for birds). This has led to the species–energy theory to explain geographical DIVERSITY patterns. The theory assumes that the likelihood of occurrence of a species increases and the probability of extinction decreases as mean ABUNDANCE increases. Productivity of a habitat determines the number of individuals per unit area; thus species diversity should increase with increasing productivity.

However, latitudinal diversity patterns in terrestrial plants cannot be explained by the theory. The relationship is stronger when based on higher taxonomic levels (order, class) than lower ones (genus, family) and better for animals than plants. On local scales, for a wide range of taxa in a wide range of habitats, the relationship between richness and productivity seems to be humped, peaking at some intermediate level of productivity. Similar unimodal relationships are found for terrestrial plants and PHYTOPLANKTON in enrichment experiments. Also, some highly productive habitats like SALT MARSHES have few species and when productivity is accompanied by a decrease in resource variety, as in polluted rivers, species richness actually declines. Productivity and the rate of flow of energy through the ECOSYSTEM therefore does not seem to have a simple relationship with species diversity. [P.S.G.]

energy-flow hypothesis

1 The energy-flow hypothesis was proposed to explain why in ELTONIAN PYRAMIDS of COMMUNITY structure there is a progressive reduction in PRODUCTIVITY in successively higher trophic levels. The hypothesis asserts that this is a consequence of the second law of thermodynamics, which states that changes in energy state are always accompanied by a loss of energy from the system as heat. Thus wherever assimilated energy is built into new BIOMASS there will always be energy expended in maintaining the basic metabolism and fuelling the activities of the animal, which will be lost as respiratory energy.

The hypothesis is often formulated as a visual model of energy flow through the community, graphically depicting respiratory heat losses from the system as energy is transferred from one TROPHIC LEVEL to the next.

2 Another, related version of the energy-flow hypothesis is that the progressive loss of energy at each trophic level along a FOOD CHAIN will put a limit on the number of links in a food chain. However, this is now not without dispute. [M.H.]

energy, laws of thermodynamics When energy is converted or transformed from one form to another the processes follow rigorous physical laws, called the laws of thermodynamics.

The first law of thermodynamics concerns the

conservation of energy and states that 'energy may be transformed from one form into another but is neither created nor destroyed'.

The second law of thermodynamics is concerned with the conversion of other forms of energy to heat and states 'processes involving energy transformations will not occur spontaneously unless there is a degradation of energy from a non-random to a random form'. The conversion of energy from more ordered to more disordered form is often referred to as the production of entropy.

This law underpins the universal observation that ecological efficiencies (the ratio of PRODUCTION by one trophic unit to the PRODUCTIVITY of the trophic unit below it) are never 100%. In ecological systems the heat lost during energy transformation is usually measured as RESPIRATION. [M.H.]

entropy *See* ENERGY, LAWS OF THERMODYNAMICS.

environment This word from the French *environer*, to encircle or surround, describes the supporting matrices of life: water, earth, ATMOSPHERE and CLIMATE. [P.C.]

environmental (ecological) economics Economics is the study of how people use limited RESOURCES to supply their unlimited needs. It is therefore concerned with trade-offs and social preferences. It can be applied internationally to trade and its consequences (macroeconomics) or to small units such as businesses and households (microeconomics). It is concerned with judging value (social preferences) in market-places by reference to willingness to pay. Until recently, resources from the environment were considered external to these processes (as externalities); this included raw materials and space available for dumping waste. For a long time, however, economists have recognized that environmental resources are finite and hence should be taken into account in terms of competing human needs (e.g. Thomas Malthus in 1797 published *Essay on the Principles of Population* that made this point). Environmental entities should therefore be valued so that they can be taken into account in the overall scheme of human economics. This internalization is an important requirement in the defining of SUSTAINABLE DEVELOPMENT. Valuing environmental entities and taking these into account in the development of policies is referred to as environmental economics. [P.C.]

environmental ethics and conservation The field of moral philosophy that deals with human obligations and duties to non-human organisms and to the natural world. At the elementary level, environmental ethics is a simple extension of human ethical systems to other living beings. This extension may be based on utilitarian concepts of pleasure and pain, which can be used to extend moral regard to sentient animals. A further extension of human moral duties may be based on the thought that all life is worthy of moral regard. This biocentric extension of human morality affords limited rights to plants and other non-sentient organisms as well as sentient animals. Further still, ethical regard may involve abstract concepts of natural process and biological relationships. This evolutionary ecological, or ecocentric, concept of ethics assigns moral conditions to the actions of humans as they affect the natural environment. [S.J.B.]

environmental impact assessment (EIA) This is generally used to refer to the evaluation of effects likely to arise from a major project, such as the construction of a dam or a power-station. However, it would also apply to smaller projects, such as the construction of a new factory or of new plant within an existing industrial operation. [P.C.]

environmental index Numerical indicator of environmental impact. [P.C.]

environmental politics Broad term for involvement of environmental issues in political processes. [P.C.]

environmental quality objective (EQO) In the UK, used to define the state of environments in terms of ecological/human health goals, for example whether water is suitable for fish stocks, suitable for drinking, suitable for bathing. Elsewhere, often synonymous with ENVIRONMENTAL QUALITY STANDARD. Usually mandatory. [P.C.]

environmental quality standard (EQS) Concentration of a substance that should not be exceeded in an environment if HARM (to humans and/or ecosystems) is to be avoided. [P.C.]

environmental science Literally, scientific endeavour applied in describing and understanding the ENVIRONMENT. It has been equated with 'earth science', namely the study of the ATMOSPHERE, land, oceans and fresh waters and the biogeochemical fluxes within them. The emphasis has been on the physics and chemistry of these processes, but increasingly the importance of biological/ecological systems is being recognized and the interface with human society is also clearly important, leading to interaction with social sciences and economics. This, then, is a broad multidisciplinary/interdisciplinary venture that plays an important part in understanding and hence mitigating our impact on natural systems and processes. [P.C.]

environmental tolerance Response of a genotype's mean FITNESS across an environmental gradient. [W.G.]

environmental toxicology Study of how human health is impacted by toxic substances through environmental exposure (usually excluding home and work-place). Sometimes used more broadly and synonymously with ECOTOXICOLOGY. [P.C.]

environmental utilization space or ecospace The capacity of the BIOSPHERE's environmental functions to support human economic activities, sometimes defined at a national or PER CAPITA level according to a 'global fair shares' principle. [P.C.]

environmental value In QUANTITATIVE GENETICS, the overall influence of an environment is quantified as the mean phenotypic value of all (relevant or investigated) GENOTYPES in that environment. [G.D.J.]

environmentalism Ideology that protection of, and respect for, environment should influence all that we do and that this is (usually) in contrast to the industrially driven, capitalist western societies. [P.C.]

environmentally friendly Shorthand for any action, industrial process or product that have, or are intended to have, reduced impact on the environment. [P.C.]

environmentally sensitive area Area designated under law as being particularly desirable to conserve, protect or enhance, for example by the adoption of particular agricultural methods. *See also* SITE OF SPECIAL SCIENTIFIC INTEREST. [P.C.]

ephemeral An organism capable of completing one or more generations in a calendar year. Ephemeral species are those that depend on rare and unpredictable environmental conditions for growth and reproduction, they tend to exhibit either fractional DORMANCY (i.e. some propagules enter long-term dormancy while others are capable of immediate germination under suitable conditions of moisture, temperature, daylength and light quality) or they possess adaptations for long-distance dispersal (often by wind), but they seldom exhibit both traits. [M.J.C. & R.H.S.]

ephemeral habitats Habitats that are transitory or short-lived. [R.H.S.]

epibenthic Living on the surface of (usually) the SEA bottom (although strictly it could also refer to freshwater systems). [V.F.]

epibiont An organism of restricted RANGE whose DISTRIBUTION has contracted from a formerly more extensive range. [P.D.M.]

epidemic An increase in the number of cases of a disease per unit time (incidence) above that expected in a defined population over a period less than the life expectancy of the HOST. Disease may be continuously present in a host population when it is referred to as endemic. An endemic infection may show a pattern of recurrent epidemics. A global epidemic is referred to as a pandemic. [G.F.M.]

epidemiology The study of patterns of disease (infectious and non-infectious) and health states in populations. Can be applied to both human and wildlife populations. [G.F.M.]

epifauna Animals that live on the bottom of aquatic environments (benthic fauna) and on the substrate surface, as opposed to living within the substrate (infauna) or resident in burrows. [J.L.]

epiflora *See* BENTHOS.

epigean Above ground. Usually refers to seed germination, for example in which cotyledons are carried above the soil. Cf. hypogean, below ground. [P.C.]

epigenetics The study of the interactions of gene products, both among themselves and with their surrounding environment, during development. Often misused to refer to non-genetic aspects of development. [P.A.]

epilithic Growing upon the surface of a rock or stone. [P.D.M.]

epiphyte An organism that grows upon the surface of a plant, using its HOST only for support. Usually used of plants growing upon trees and shrubs, but it can also be used of MICROFAUNA in aquatic situations. [P.D.M.]

episodic pollution Emissions and EFFLUENTS that come in bursts, for example from stormwater discharges. [P.C.]

epistasis Refers to the interaction among non-homologous genes. [I.O.]

epizoic *See* BENTHIC HABITAT CLASSIFICATION.

epizootic An EPIDEMIC in a non-human animal population. [G.F.M.]

equilibrium, communities Equilibrium of communities is considered in two rather different contexts: indefinite COEXISTENCE of species in mixture, and ability of communities to return to a certain condition after DISTURBANCE. In the first case, equilibrium is said to occur where the species present can coexist indefinitely as a result of NICHE differentiation.

A community's ability to return to a given state after disturbance is termed 'resilience'. Resilience may occur *in situ* (from root sprouts and the SEED BANK) or by migration (movement back into an area where the whole plant cover has been destroyed). *See also* RESILIENCE. [P.J.G.]

equilibrium, populations Refers to a POPULATION, the size of which is not changing through time, or the characteristics of which are not changing through time. A population's size is at equilibrium when births and IMMIGRATION balance deaths and EMIGRATION. [S.N.W.]

equitability In no ecological community are all species equally common. Instead, some species will be very abundant, others moderately common, with the remainder infrequent or rare. Equitability (sometimes also termed 'evenness') is a measure of the extent to which species are equally represented in a community. [A.E.M.]

erosion and topsoil loss Loss of topsoil as a result of physical removal by WIND and water, and physical, chemical and biological changes. [P.C.]

essential elements There are 17 elements essential for plant GROWTH. From the air and water, carbon, hydrogen and oxygen are derived. From the SOIL the following are obtained: nitrogen, phosphorus, potassium, calcium, magnesium, sulphur and, in trace quantities, iron, manganese, boron, molybdenum, zinc, chlorine and cobalt. The last seven are termed 'MICRONUTRIENTS' or 'trace elements'. A few other elements are apparently needed as micronutrients by some plant or animal species:

sodium, fluorine, iodine, silicon, strontium and barium. [J.G.]

establishment A technical term used in different ways in different subject areas.

1 Metapopulation dynamics. Following colonization of a patch, there are several possibilities: all the propagules might die; the propagules might produce individuals that live for a long time, but these individuals never reproduce; or the colonists may reproduce, giving rise to an expanding new population and producing migrants for the colonization of new patches. Only this latter case constitutes establishment. Species where the colonists survive but do not reproduce are called CASUAL SPECIES.

2 ISLAND BIOGEOGRAPHY. Establishment is the formation of a breeding population of a species new to an island. Species arrive on islands at a rate inversely proportional to the number of species already established and directly proportional to the size of the species pool in the source region from which the colonists originate.

3 POPULATION DYNAMICS. The formation of a local breeding population (*cf.* RECRUITMENT). An established population must exhibit the ability to increase when rare; technically, this property requires that $dN/dt > 0$ when N is small (the invasion criterion; *see also* ALLEE EFFECT).

[M.J.C.]

estimated (predicted) environmental concentration Predicted concentration of XENIOBIOTIC in an environmental compartment based on information on patterns and amounts of production, use and disposal, and physicochemical properties of the chemicals. [P.C.]

estuary A semi-enclosed body of water that has a free connection with the open SEA and within which seawater is diluted measurably with FRESH WATER derived from land drainage. [V.F.]

eukaryote Organism whose genome consists of chromosomes, in which DNA is complexed with histones, surrounded by a membrane to form a nucleus. Consists of protistans, algae, fungi, plants and animals. *Cf.* PROKARYOTE. [P.C.]

Euler–Lotka equation Also known as the Euler equation, the Lotka equation or the characteristic equation, this allows calculation of the intrinsic rate of population growth, r, from knowledge of the life histories of the individuals in the population. Suppose that individuals breed at ages $t_1, t_2, t_3, \ldots$ producing $n_1, n_2, n_3 \ldots$ female offspring and that survivorship from birth to ages $t_1, t_2, t_3, \ldots$ is $l_1, l_2, l_3, \ldots$, then the equation is:

$$1 = n_1 l_1 e^{-rt_1} + n_2 l_2 e^{-rt_2} + n_3 l_3 e^{-rt_3} + \cdots$$

If the individuals breed at ages 1, 2, 3,... then the equation can be written:

$$1 = \sum_{x=1}^{\infty} n_x l_x e^{-rx}$$

The value of r calculated from the equation can be used to forecast future POPULATION SIZE provided that the life histories and the relative numbers of individuals in each AGE CLASS do not change, i.e. the population has a STABLE AGE DISTRIBUTION. *See also* LIFE-HISTORY EVOLUTION; POPULATION DYNAMICS; POPULATION REGULATION. [R.M.S. & D.A.R.]

eulittoral zone Generally also called the LITTORAL ZONE (although sometimes classified as a subdivision of this zone) or INTERTIDAL ZONE. [V.F.]

euphotic zone Also called the photic or epipelagic zone. Refers to the lighted part of the OCEAN in which PRIMARY PRODUCTIVITY occurs. It extends from the surface to the depth at which PHOTOSYNTHESIS is no longer possible due to a lack of light. [V.F.]

eury- From the Greek *eurus* meaning wide. Applied as prefix meaning wide range of tolerance for some environmental factors, for example eurybathic for hydrostatic pressure, eurythermal for TEMPERATURE, euryhaline for SALINITY. *Cf.* STENO-. [P.C. & V.F.]

eutrophication Biological effects of an increase in plant NUTRIENTS (usually nitrogen and phosphorus, but sometimes silicon, potassium, calcium, iron or manganese) on aquatic systems. [P.C.]

evenness *See* EQUITABILITY.

evergreen Evergreen leaves and other plant organs are not seasonally deciduous and have a photosynthetic life of a few to many years, thus allowing PHOTOSYNTHESIS at any time of year when temperature or water is not limiting. [J.R.E.]

evolution Evolution is a comprehensive framework that combines with physics and chemistry to explain all biological phenomena. It is the only part of biology containing principles not implicit in physics and chemistry. It has three major principles: NATURAL SELECTION, inheritance and history; and one fundamental property: selection acts on organisms but the response to selection occurs in stored information. The first principle, natural selection, is a great law of science, the only mechanism known that maintains and increases the complexity of organisms. Variation among organisms in reproductive success produces natural selection. Populations of organisms respond to selection when some of that variation is genetically based; this happens in stored information. The result is a genetically based change in the phenotypic design of offspring from the more reproductively successful parents. [S.C.S.]

evolutionarily stable strategy (ESS) Often referred to as 'evolutionary' but strictly should be 'evolutionarily', the point being that EVOLUTION is not 'goal-directed' and so should not be thought of as having strategies. Some complex ADAPTATIONS are more likely to succeed and persist in populations than others and these are evolutionarily stable. [P.C.]

evolutionary optimization A research programme that aims at interpretation of structures, behaviours and other properties of organisms in terms of the contribution they make to Darwinian FITNESS. It is based on the assumption that NATURAL SELECTION has led to the EVOLUTION of PHENOTYPES that confer the highest fitness possible under given circumstances. An optimization study makes assumptions about the functional CONSTRAINTS and TRADE-OFFS that delimit the set of possible phenotypes, about the optimization criterion (fitness measure) and about the pattern of inheritance. [T.J.K.]

evolutionary rate The speed of evolutionary change. For characters whose measurement has changed from x_1 to x_2 during a time-interval of length Δt (in years or millions of years), the evolutionary rate, r, can be calculated as $r = \ln(x_2) - \ln(x_1)/\Delta t$ (in units of darwin). Logarithms are taken to make the evolutionary rate independent of scale (i.e. the size of an organism). [W.W.W.]

evolutionary trends An evolutionary change in a given direction that is maintained for a long time. Evolutionary trends such as a general increase in SIZE (COPE'S RULE) are commonly observed in the FOSSIL RECORD. They can be seen either as a persistent character change within a single LINEAGE or as a change in the average over several related lineages. [W.W.W.]

e_x In a LIFE TABLE the expectancy of further life of the individuals remaining alive at age or stage x. Calculated as T_x/l_x where:

$$T_x = \sum_x^{\infty}\left(\frac{l_x + l_{x+1}}{2}\right)$$

[A.J.D.]

***ex situ* conservation** *Ex situ*, or off-site, conservation is the practice of maintaining individuals or genetic material of species at risk at a site removed from their natural habitats. For plants, methods include storing seed (SEED BANKS), storing pollen, growing plants in cultivation (field gene banks), maintaining plant material as undifferentiated tissue (tissue culture or micropropagation), grafting on to stock plants, or even storing genetic material directly such as in gene banks of DNA. *Ex situ* methods for animals include maintaining breeding populations in captivity, which may be distributed among many geographically dispersed institutions, or storing ova and sperm. [E.O.G. & L.R.M.]

***ex situ* genetic reserves** Conservation of species (and often SUBSPECIES) is increasingly focused on botanic gardens, germplasm centres, microbial culture and gene library collections. Together with zoos, they constitute *ex situ* as opposed to *in situ* conservation (*see IN SITU* GENETIC RESERVES), the latter more the provenance of nature reserves or natural habitats protected by conservation legislation. [G.A.F.H.]

exact compensation Density-dependent MORTALITY, birth or growth rate that leads to a return to population equilibrium in one generation, following DISTURBANCE from equilibrium. Exact COMPENSATION is sometimes said to be the ecological consequence of pure CONTEST COMPETITION. Density-dependent mortality is detected by plotting $k = \log_{10}$(initial density/final density) against $\log_{10}$(initial density); if the slope $b = 1$, compensation is exact. *See also* OVERCOMPENSATION; POPULATION REGULATION; UNDERCOMPENSATION. [R.H.S.]

exaptation A word constructed, like 'adaptation', from an adjective plus 'apt', meaning fitted. An exaptation is a character currently selected that arose in some way unrelated to its present function. [L.M.C.]

exclosure An exclosure is any area from which animals are excluded. Exclosures are typically used to prevent the entry of GRAZING mammals (wild or domesticated) into an area that is protected for scientific or ecological management purposes. [D.A.F.]

exotic and invasive species Species of plants and animals that have been transported, with human aid, beyond the limits of their native GEOGRAPHIC RANGES and which continue to expand their distributions by displacing species indigenous to the invaded areas. [L.L.L.]

experimental design Ensuring that causes can be associated with effects by carefully controlling all appropriate variables, having adequate replication and applying sound statistics. [B.S.]

experimental ecology Ecology carried out according to the experimental ideal. *See also* EXPERIMENTAL DESIGN. [P.C.]

experimental lake area (ELA) Area containing a number of lake basins in north-western Ontario, Canada that have been subjected to comprehensive MANIPULATION to explore ECOSYSTEM responses to EUTROPHICATION, ACIDIFICATION and POLLUTION. [P.C.]

exploitation competition COMPETITION that does not involve direct interaction between individuals, i.e. competition is passive because there is no INTERFERENCE. Under exploitation competition, individual FITNESS is determined by the level of different RESOURCES that remain in the zone where competition takes place after resources have been exploited by others. [R.H.S.]

explosive breeding A phenomenon most commonly found in anurans (frogs and toads) and some insects in which all mating occurs during a very short breeding season, of the order of 1 or 2 days. [J.C.D.]

exponential distribution A continuous PROBABILITY DISTRIBUTION that is commonly used to describe a variety of ecological processes, such as MORTALITY rates and DISPERSAL distributions. In statistical analyses, GENERALIZED LINEAR MODELS rely on the exponential distribution or the family of distributions related to it. The exponential distribution arises when the probability of some discrete event occurring at any moment in time is constant. [R.P.F.]

exponential growth Unlimited population growth or population growth with no COMPETITION for RESOURCES. Exponential growth is only possible in the short term while conditions for population increase are favourable and effectively constant, for example during the log phase of bacterial population growth. Characterized by a constant average rate of increase per individual or POPULATION GROWTH RATE *r* (also known as the INTRINSIC RATE OF INCREASE). [R.H.S.]

exported production The fraction of primary PRODUCTION that is transported from where it was produced to somewhere else; very often refers to transport from the EUPHOTIC ZONE to the DEEP SEA. [V.F.]

extinction crisis, current global A total of about 490 animal extinctions and 580 plant extinctions have been recorded globally since 1600. The number of extinctions has increased dramatically in the last 100 years. There has been a preponderance of extinctions on islands compared to continents. For example, 75% of recorded animal extinctions since 1600 have been of species inhabiting islands, even though islands support a small fraction of the number of animal species found on continents. The apparent decline in extinctions in the last 30 years to 1990 is at least partly due to the time-lag between extinction events and their detection and recording. A number of species are likely to have become extinct recently without yet having been recorded as such. It is also possible that conservation efforts over the last 30 years have slowed the rate of extinctions. [M.A.B.]

extinction in the fossil record Most species that have ever existed are now extinct. Today's biotic diversity constitutes perhaps between 1 and 10% of the total that has ever been. Many species were not eliminated in the sense that we think of extinction today. Rather, NATURAL SELECTION and MUTATION transformed many species, giving the appearance of losses and gains in the fossil record. Others underwent change following geographic subdivision of an ancestral population or following hybridization between different species after DISTURBANCE or DISPERSAL.

The average lifetime of species in most taxa is a few million years. For example, invertebrates average 11 million years, marine invertebrates average 5–10 million years and mammals average 1 million years. Most species within a major taxonomic group tend to be short-lived relative to the average lifetime for species in the TAXON, while a few species persist for much longer.

Extinctions in the fossil record are generally typified as falling within one of two regimes: background periods and MASS EXTINCTION events. Mass extinctions are substantial BIODIVERSITY losses that are global in extent, taxonomically broad and rapid relative to the average duration of the taxa involved. The background rate of extinction is the average rate of loss of species globally in geological time, outside the periods of mass extinction. If there are 10 million species currently in existence and the average lifespan of a species is 5–10 million years, then very roughly the background extinction rate is of the order of one or two species per year globally. The background rate over the last 600 million years has been punctuated by five mass extinction events, each associated with relatively severe environmental change. Despite the severity and extent of mass extinction events, it is likely that more than 90% of all extinctions occurred in background times. [M.A.B.]

extinction modelling Extinction modelling is the process of building conceptual, statistical and mathematical models to summarize our understanding of the dynamics of extinction and to predict patterns of extinction.

At the level of communities within regions, the theory of ISLAND BIOGEOGRAPHY is sometimes used to predict numbers of extinctions. At equilibrium between IMMIGRATION and extinction, the number of species in a HABITAT patch is estimated by $S = cA^z$ where A is area. The proportional change in the number of species, ΔS, is given by:

$$\Delta S = (S_1/S_0) = 1-(A_1/A_0)^z$$

where ΔS is a function of the remaining habitat. Normally, the value of z is taken to be about 0.25 with a range of $0.15 < z < 0.35$.

The likelihood of extinction of species has been estimated by developing statistical models of local extinction (extirpation) based on species characteristics. Some factors important to mammals and birds include POPULATION SIZE and range, body size, social structure and the ability to use disturbed habitat.

Extinction risks for individual species may be estimated by developing dynamic mathematical models that include stochastic elements representing sources of UNCERTAINTY, a process known as population viability analysis. It may be used to explore how management activities influence the systematic decline or PROBABILITY of extinction of a species. The probability of extinction is usually estimated by MONTE CARLO SIMULATION of the factors that govern the chances of persistence of the species including demographic parameters (survivorships and fecundities), density-dependent REGULATION, environmental variation, spatial correlation of environmental variation, DISPERSAL, changes in the status of habitat and the impact of infrequent, catastrophic events such as fire. Appropriate model structure depends on consideration of the ecology of the species, data availability and requirements, and an understanding of management needs. *See also* ENDANGERED SPECIES; EXTINCTION CRISIS, CURRENT GLOBAL; EXTINCTION IN THE FOSSIL RECORD; MASS EXTINCTION. [M.A.B.]

extra-pair copulation (EPC) A copulation between a mated individual and an individual of the opposite sex to which it is not socially bonded. [A.P.M.]

extrinsic mortality Mortality caused by external factors such as predation, accident (e.g. fire, drowning), famine or infection. The term is used to distinguish these kinds of death from those associated with INTRINSIC MORTALITY. [T.B.L.K.]

F

F-statistics POPULATION GENETICS' parameters and statistical tools used to study the partitioning of variance of a given TRAIT (frequency of an allele at a locus or value of a quantitative character) into hierarchical effects. For instance, *F*-statistics are often used to describe the variance of allele frequencies among populations, among subpopulations within populations, among individuals within subpopulations, and among ALLELES within individuals. [I.O.]

***F* test** A statistical procedure that tests the NULL HYPOTHESIS (H_0) that two SAMPLE variances are from POPULATIONS with equal variances (H_0: $\sigma_1^2 = \sigma_2^2$). One variance is divided by the other to produce a ratio that will equal one if the two variances are equal. This ratio (the variance ratio) has been called *F*, in honour of R.A. Fisher (*see* CHARACTERS IN ECOLOGY), an English statistical biologist, and its formula is simply:

$$F = \frac{s_1^2}{s_2^2}$$

If the alternative hypothesis (H_1) is: $\sigma_1^2 \neq \sigma_2^2$ it is a TWO-TAILED TEST. This is the case when performing this test before carrying out a *z* test or *t* TEST to see if the two variances are homogeneous. The test is two-tailed because we have no a priori reason to expect one particular variance to be greater. If the alternative hypothesis is H_1: $\sigma_1^2 > \sigma_2^2$ it is a ONE-TAILED TEST. This is the most frequent use of the *F* test, in ANALYSIS OF VARIANCE.

STATISTICAL TABLES of *F* show DEGREES OF FREEDOM for both numerator (ν_1) (usually columns) and denominator (ν_2) (usually rows). Therefore, unlike other statistical tables (t, r, χ^2), levels of PROBABILITY are shown in separate tables. Some statistical books show separate 5% tables for both one-tailed and two-tailed tests, while others show only one-tailed tables, but have both 5% and $2^1/_2$% tables. The latter should be used as 5% tables in a two-tailed test. [B.S.]

facilitation Facilitation occurs during SUCCESSION when the chance (or speed) of ESTABLISHMENT of a given species is increased by the prior occupation of the site by another species. [P.J.G.]

facultative A facultative factor is one which does not have to be present in order for an organism to survive in a particular environment. [R.C.]

facultative annual An ANNUAL plant species in which not all individuals complete their LIFE CYCLES within the year of birth or germination. [P.O.]

faecal pellets Defecated material (usually used in reference to aquatic invertebrates) that forms compacted aggregations of particles released by egestion. [V.F.]

faunal region The term faunal region is used for land and freshwater vertebrates. Six main faunal regions are now generally accepted. They were first proposed in 1858 by P.L. Sclater and were confirmed by A.R. Wallace in 1876 for vertebrates in general and for some invertebrates also. The accepted system of continental faunal regions is as follows:

1 Ethiopian region: Africa south of the Sahara, with part of southern Arabia.

2 Oriental region: tropical Asia, with associated continental islands.

3 Palaearctic region: Eurasia north of the tropics, with the northern limit of Africa.

4 Nearctic region: North America excepting tropical Mexico.

5 Neotropical region: South and Central America including tropical Mexico.

6 Australian region: Australia, with New Guinea, etc.

[A.H.]

favourability The contrast is sometimes made between favourable (benign) and unfavourable (hostile or stressful) environments. The notion of environmental favourability (*see* GRIME'S TRIANGLE) is problematical, since all environments are hostile in one way or another. [M.J.C.]

favourableness A term sometimes used in the CLASSIFICATION of SELECTION PRESSURES. [P.C.]

fecundity A measure of the number of offspring produced by an individual during a period of time. Fecundity is sometimes used interchangeably with fertility, but fecundity is the preferred term used to describe reproductive output in ecology—fertility has other meanings; for example, the potential of an animal to produce offspring, or the ability of soil to support plant growth. [R.H.S.]

fecundity schedule A tabulation of the pattern of offspring production (NATALITY) of individuals with respect to their age. Such a schedule, also called a FECUNDITY table, is frequently included in a LIFE TABLE, and contains F_x (total number of offspring produced by the population at age x), $m_x = F_x / a_x$ (age-specific fecundity or mean number of offspring produced by each individual at age x), and $V_x = l_x m_x$ (the number of offspring produced at age x per individual in the initial COHORT). (In fecundity schedules, natalities are expressed in relation to females alone by halving a_x and l_x.) The BASIC REPRODUCTIVE RATE, R_0, is computed by summing the values of V_x. [A.J.D.]

feedback Information regarding a control mechanism or controlled system that enables actual performance to be compared with target performance. Performance can then be adjusted either positively or negatively, corresponding to, respectively, either positive or NEGATIVE FEEDBACK. Feedback can involve physical feedback loops, as with thermostats, reflex responses and management systems. Alternatively, feedback can be apparent, for example when liquid escapes from a leaking tank the rate of outflow reduces as volume reduces, giving the impression that there is feedback control on rate of loss, but there is no feedback hardware here. Similarly, as POPULATION SIZE reduces under predation, there may be more food per individual, which causes increased FECUNDITY and RECRUITMENT, so reducing population decline. Again this might appear to be feedback, but there is no active control here that helps keep population to any goal DENSITY in the same sense as a thermostat actively controls temperature to a goal temperature. *See also* SYSTEMS ECOLOGY. [P.C.]

feeding Feeding includes all behaviour directed towards acquiring, manipulating and consuming food. It is an essential activity for heterotrophic organisms (including non-photosynthetic parasitic plants), providing them with the energy and NUTRIENTS required for their GROWTH, maintenance and reproduction. It is also found in facultatively autotrophic organisms (mostly unicellular organisms, which are autotrophic in the presence of light but which switch to heterotrophy in its absence), in autotrophic hemiparasitic plants (which obtain soil nutrients and water by parasitizing the roots of their plant HOSTS), and in insectivorous plants (which obtain amino acids from their prey). [M.M.]

feeding interference Reduction of food intake caused by the presence of other individuals (normally conspecific). [J.E.]

feeding types, classification Food occurs in many forms. The supply is nevertheless limited, so that NATURAL SELECTION will have tended to promote diversification in FEEDING habits; this should have resulted in the complete exploitation of all potential foods on this planet. A number of schemes of CLASSIFICATION have been used to give some order to the large quantity of information available on feeding. Three main ones are listed below. *See also* FOOD SELECTION; OPTIMAL FORAGING THEORY.

1 *Particle size.*

• Microphagous feeders. These are animals that ingest small particles (<1 mm diameter). They may feed on particles suspended in solution (suspension-feeders) or deposited on the substratum (DEPOSIT-FEEDERS).

• Macrophagous feeders. These can be divided into those animals that seize and swallow only and those that use some kind of preingestion processing mechanism.

• Fluid-feeders. These use NUTRIENTS that are dissolved in aqueous solutions and take them by soaking, or sucking, or piercing and sucking.

2 *Functional feeding groups.* A related classification has been developed for benthic aquatic invertebrates where the feeding types are referred to as functional feeding groups. These are:

• shredders—feed on COARSE-PARTICULATE ORGANIC MATTER (CPOM) or live macrophytes;

• collectors—feed on FINE-PARTICULATE ORGANIC MATTER (FPOM) either by filtering or gathering;

• scrapers—feed on attached PERIPHYTON;

• piercers—feed on macroalgae by piercing individual cells;

• predators—feed on other animals.

3 *Trophic status.* Another way of classifying foods is in terms of their own trophic status (i.e. into AUTOTROPHS, which can manufacture organic foods from inorganic precursors, and HETEROTROPHS, which cannot) and their condition before being eaten (living or dead).

[P.C.]

female preference The phenomenon whereby a characteristic of a female makes her prefer to mate with one male rather than another on the basis of differences in a male TRAIT. If there is genetic variation in the population for the extent to which the male trait is developed, either natural or (inter) SEXUAL SELECTION will result in the trait developing in the direction of the female preference, resulting in sexual dimorphism if the trait is a physical characteristic. [M.M.]

femtoplankton Planktonic organisms of size in the range 0.02–0.2 μm, i.e. viruses. [V.F.]

fen A RHEOTROPHIC wetland ecosystem, dominated by herbaceous plants, in which the WATER TABLE approximately coincides with the soil surface during drier times of year. [P.D.M.]

feral Living in a wild or NATURAL state. Used especially for domesticated animals and plants that become wild. [P.C.]

fertility *See* FECUNDITY.

fertility schedule *See* FECUNDITY SCHEDULE.

fertilization The process whereby the nucleus of a male GAMETE (e.g. SPERMATOZOA) fuses with that of a female gamete (e.g. an ovum) to form a cell (zygote) with a single nucleus. In animals, fertilization of EGGS can occur outside or within the maternal body. Both modes are often found within a phylum. [R.N.H.]

fertilizers Materials added to the SOIL to increase the PRODUCTIVITY of farmland or forest by supplying chemical elements that are at suboptimal levels for plant growth. [G.R.]

field A term often used to describe NATURAL sites or systems, for example field experiments are carried out in the open as opposed to in laboratories. [P.C.]

fine-particulate organic matter (FPOM) Plant DETRITUS in aquatic systems between *c.* 0.45 μm and 1 mm. *Cf.* COARSE-PARTICULATE ORGANIC MATTER. [P.C.]

fish farming The culture of fish (often used broadly to include both fish and molluscs!), in indoor/outdoor tanks or enclosures in natural water bodies, for human consumption. *See also* AQUACULTURE. [P.C.]

fish stocks The standing stock is the total mass of fish present at a given time in a fishery. [V.F.]

Fisher's exact test This is a test for ASSOCIATION in a 2 × 2 CONTINGENCY TABLE. It is called an exact test because the exact probabilities of every similar 2 × 2 table, with the same marginal totals, are calculated. It is used when expected values are less than 5 and the CHI-SQUARED TEST (χ^2) becomes approximate. [B.S.]

Fisher's fundamental theorem The theorem, as stated by R.A. Fisher (*see* CHARACTERS IN ECOLOGY), that 'the rate of increase in FITNESS of any organism at any time is equal to its genetic variance in fitness at that time'. It is often misunderstood to mean that selection will lead continually to an increase in average fitness. Numerous authors have shown that this interpretation of the theorem is incorrect since there are many circumstances (e.g. presence of EPISTASIS, linkage disequilibrium) in which average fitness does not increase. The change in fitness can be partitioned into two components, one due to NATURAL SELECTION and one due to environmental change. Fisher's theorem only refers to the first component and means that the change in this component of fitness is equal to the additive variance in fitness. Changes in environmental conditions may, however, cause total fitness to decline. [D.A.R.]

Fisher's sex-ratio theory A theory proposed by R.A. Fisher in 1930 to explain the EVOLUTION of stable SEX RATIOS. The theory predicts that a stable ratio of males and females in a population will be that at which the parental expenditure on males and females is equal. [V.F.]

fission A mode of ASEXUAL REPRODUCTION. [P.C.]

fitness NATURAL SELECTION on a PHENOTYPE or GENOTYPE involves a quantity known by the technical term 'fitness'. 'Fitness' indicates a short-term measure of expected reproductive success. Fitness is not a property of an individual: fitness summarizes a model for genotypic DEMOGRAPHY that makes certain assumptions about the population and the genotypic and individual interactions within it. In order to arrive at genotypic fitness in a population, the genotypic differences between fitness components such as age-dependent viability, age-dependent FECUNDITY and age at maturity should be known; one should also know whether individuals can be regarded as independent in survival and reproduction, or whether they interact. [G.D.J.]

fitness profile The mathematical function, also called fitness function, describing the relationship between a QUANTITATIVE TRAIT and FITNESS. Fitness profiles reflect the direction and strength of NATURAL SELECTION. Monotonic fitness profiles correspond to DIRECTIONAL SELECTION; those with an intermediate optimum within the population range of variability describe STABILIZING SELECTION. Fitness profiles with an intermediate optimum are often asymmetric, which results in higher fitness costs of deviating in one direction from the optimum than in the other. In these circumstances selection may shift the population mean off the optimum in the direction of slower decline of fitness. [T.J.K.]

fixed-quota harvesting A HARVESTING strategy in which harvesting activity is regulated by a fixed catch level. [J.B.]

flagship species Popular, and usually charismatic, species that provide a symbol and focus for conservation awareness and action, for example the panda and dolphins. [P.C.]

floral province A BIOGEOGRAPHICAL UNIT based on the component plants. [W.G.C.]

Floras Floras are publications giving a taxonomic (systematic) treatment of plants of a particular geographical area, which may be a country, region or smaller area. Floras may cover one or several major plant groups; most include vascular plants only. Treatment may be general but more detailed for small areas, with full descriptions of taxa and information, for example on habitat, distribution and ecology. Many Floras have keys for identification; not all are illustrated. [A.J.W.]

floristics The totality of plant species found in a given site. [P.D.M.]

fluctuations Variations over time, for example in the size of a population or in environmental variables such as temperature, salinity, pH, etc. [V.F.]

food chain A series of organisms linked by their FEEDING relationships. Each link in the chain feeds on, and obtains energy from, the preceding link and provides food and energy for the following link. The number of links in the chain is limited: for example primary producer → herbivore → primary carnivore → secondary carnivore. [M.H.]

food selection Using the net rate of energy gain as currency, models predicting optimal diets (i.e. food selection) have been applied to all major trophic categories except endoparasites. [R.N.H.]

food web Food webs are ways of describing the structure of ecological communities, and the trophic interactions which occur between the component units of a COMMUNITY. They illustrate that FOOD CHAINS are not isolated sequences but are interconnected with each other. [M.H.]

foraging, economics of The profitability of behaviours for acquiring energy and NUTRIENTS. Examples of decisions are selection of types of food (diet choice; *see* FOOD SELECTION) and the time before moving on to a new area (patch choice). Increasing (generalizing) or decreasing (specializing) the number of prey types eaten represents a behavioural decision influencing the economics of diet choice. [J.E.]

foreshore The portion of the shore lying between the normal high- and low-water marks; the INTERTIDAL ZONE. [V.F.]

forest Nowadays, this term usually refers to a large area of tree-covered land, but including other habitats in a matrix of trees. However, in Britain, it originally referred to unclaimed land beyond the bounds of farmland and habitation (cf. 'foreign', possibly from the same root). This land was commonly wooded, but not necessarily so. [G.F.P.]

forest soil Any SOIL that has developed under a FOREST cover. [K.A.S.]

forestation The creation of FOREST conditions on an area of land which has not been forested for a long time. [J.F.B.]

fossil assemblage The fossilized remains of former plant and animal communities that are available for investigation by the palaeontologist or palaeoecologist. [M.O'C.]

fossil record Fossils are the remains or imprints of animals and plants of the geological past, naturally preserved by burial in sediments. Fossils provide a record of the EVOLUTION of life on Earth, furnish clues to past environments and provide basic data for the history of the Earth. The fossil record is used to correlate geographically separated rock strata, in part to estimate their age. Fossils are first observed in abundance in rocks from the Cambrian period. Although sedimentary rocks from the Precambrian may contain primitive fossils, these are either sparse or lack characteristic features useful for correlation. [V.F.]

founder-controlled communities 'Possession is nine-tenths of the law.' The ecological processes involved in founder control are site pre-emption and INHIBITION OF SUCCESSION. In a COMMUNITY where the species possess roughly equal DISPERSAL abilities and exhibit similar competitive abilities, then COEXISTENCE is possible and COMPETITIVE EXCLUSION need not occur if there is founder control. The dynamics of this kind of INTERSPECIFIC COMPETITION, and the necessary and sufficient conditions for coexistence, have been worked out by the use of LOTTERY MODELS. [M.J.C.]

founder effect A small population (at its smallest, one or two individuals) may be accidentally isolated; the descendants then increase in numbers to form a successful new population. In doing so, an opportunity is presented for a new coadapted GENE POOL to be formed from the subset of genes carried by the founders, augmented by those which subsequently enter the population by MUTATION. Evolution has taken a new course and the new complex may be at least as successful as, and become reproductively isolated from, the parental type. [L.M.C.]

fractal analysis Benoit Mandelbrot in his *Fractal Geometry of Nature* has demonstrated that the length of a complex structure such as a coastline depends upon the SCALE at which it is measured. At 1 km scale using an atlas we get one answer, at the scale of a human pace (1 m), walking around it, we get another (larger) answer. So geometrical objects such as this do not have one fixed dimension but a fractal dimension, D, such that the perceived length $L(\lambda)$ depends upon 'step length' of the measurement, λ, such that:

$$L(\lambda) = C\lambda^{1-D}$$

where C = constant. When $D = 1$, we have the familiar Euclidean situation in which L is independent of measuring scale.

This concept has a number of implications and applications for ecology. For example, organisms at different sizes can perceive landscape and habitat in different ways. Alternatively, abrupt changes in the fractal dimension D can signal different ecological processes operating at different scales. [P.C. & V.F.]

fragile dynamics Dynamics which change dramatically with small parameter changes, or slight changes in conditions or model structure, or when subjected to small perturbations. For example, chaotic POPULATION DYNAMICS (*see* CHAOS). [S.N.W.]

fragmentation of habitats and conservation Fragmentation of natural systems is a pervasive ecological phenomenon and a central problem in conservation and restoration ecology. As forested landscapes become fragmented, changes occur in the physical environment of forests, the ability of organisms to migrate and disperse across the landscape, and the spatial and temporal dynamics of DISTURBANCE. With increased spatial DISPERSION of suitable or accessible HABITAT, populations of some species may experience reduced success in dispersing to, or colonizing, new sites. Populations may also decline as habitat area is reduced and may become increasingly vulnerable to local extinction due to genetic, demographic or environmental stochasticity. Finally, fragmentation may favour some groups of species over others, resulting in changes in biotic composition of the forested landscape.

Consequently, fragmentation is a central issue in the conservation and management of parks and protected areas. By virtue of their fragmentation, many protected ECOSYSTEMS become modified remnants of once large and more continuous systems. There is growing concern that over time, fragmentation will cause degradation of the ecological characteristics of many protected areas. [D.A.F.]

frequency-dependent competition A situation in which the outcome of INTERSPECIFIC COMPETITION is dependent upon the frequency of the competing species. It is a stabilizing force and, although not explicitly stated, is present in all models/descriptions of competition that allow for a stable COEXISTENCE of species. *See also* POPULATION REGULATION. [B.S.]

frequency-dependent selection Frequency-dependent selection implies that the FITNESS differences underlying selection depend upon the frequencies of the GENOTYPES in the population. *See also* NATURAL SELECTION. [G.D.J.]

fresh water Water with no significant amount of dissolved salts or minerals. It generally contains <1000 mg l^{-1} of dissolved solids. *Cf.* SEA. [P.C.]

frugivore An animal specializing in FEEDING on fruits. *See also* FEEDING TYPES, CLASSIFICATION. [P.O.]

fugitive species A competitively inferior species that coexists with a competitively superior species by having better powers of DISPERSAL. The fugitive species finds suitable patches of habitat first (on average), reproduces and leaves before the competitively superior species arrives. [B.S.]

functional explanation An explanation for a characteristic of an animal or plant based on FITNESS-related performance. The alternative would be phylogenetic, a carry-over from a previous state. [L.M.C.]

functional feeding group *See* FEEDING TYPES, CLASSIFICATION.

functional group A functional group is a set of biological entities that possess common functional or structural attributes. Such groups often comprise biological species, but groupings can be made at any level of biological organization. A related concept is that of the GUILD, which groups species on the basis of shared RESOURCES. By using functional groups, the general principles which govern the functioning ecological systems may often be more easily understood. When considering predation, for example, classifying animals into true predators, grazers, parasites and parasitoids offers ways of exploring and clarifying ecological relationships that would be obscured if these functional distinctions were not made. [S.J.H.]

functional response The number of prey (HOSTS) that are consumed (attacked) by a PREDATOR (PARASITOID) as a function of host DENSITY. Three types are normally considered.

• Type 1: the number consumed increases linearly with prey density until a THRESHOLD density, beyond which a constant number is consumed. It probably applies best to animals such as filter-feeders.

• Type 2: the number consumed increases with prey density at a decelerating rate until an asymptote is reached. Models such as HOLLING'S DISC EQUATION which constant encounter rates and a finite HANDLING TIME give rise to type 2 responses.

• Type 3: the relationship between numbers consumed and prey density is sigmoid (S-shaped). A type 3 functional response occurs if predators are inefficient at detecting rare prey (perhaps because they do not develop a search image) or if they switch to attacking more common items of prey.

[H.C.J.G.]

fundamental niche (potential niche) The NICHE in the absence of competitors and predators. *See also* REALIZED NICHE. [J.G.S.]

fungicides PESTICIDES designed to eradicate or control fungal pathogens on plants, animals and materials derived from organic sources. [K.D.]

G

Gaia In Greek mythology, Gaia, or Gaea, is the goddess of the Earth. The name has been used by James Lovelock for his theory that the planet operates as a self-regulatory system—a SUPERORGANISM—to ensure that life is maintained. This does not necessarily mean that human life will be maintained.

A major problem with the Gaia theory is that it presumes that the activities of species are somehow coordinated for the common good, whereas Darwinian selection favours species that behave selfishly! Moreover, life on the planet does not appear to be as tightly coordinated and integrated as the cells, tissues and organs are within an organism. [P.C.]

gallery forest A type of FOREST vegetation found along the stream and river banks within dry SAVANNAH grasslands. GRAZING and an annual rainfall of less than 1500 mm excludes forest from the grassy plains of the savannah, but beside the watercourses the moist soils support forest. Unlike the savannah grasslands the gallery forest is relatively resistant to fire. [M.I.]

game theory Organisms, or at least the TRAITS that they carry, can be thought of as playing games for existence with each other, and game theory can be used to define the optimum moves in the game. These behavioural games are most vividly illustrated by aggressive interactions between players; when the outcome, in terms of winners (with highest FITNESS), depends upon what other individuals do. [P.C.]

gamete Sexual reproductive PROPAGULE that is haploid (*see* HAPLOIDY) and that generally requires to fuse with another in FERTILIZATION to initiate development (*cf.* PARTHENOGENESIS). *See also* ANISOGAMY; EGGS; OVA; SPERMATAZOA. [P.C.]

gamete-order hypothesis A hypothesis, also called order of GAMETE release hypothesis, or cruel bind hypothesis, to explain which sex provides PARENTAL CARE in species in which only one parent attends the young. In species with internal FERTILIZATION, such as mammals, since females could not desert their young immediately after mating, males could abandon the female first. This leaves her in a 'cruel bind'—abandon the young and let them die, or remain with the young at a cost to future reproduction. This, it was argued, would select for female parental care. Conversely, with external fertilization, typical of most teleost fish, females were thought to be able to produce gametes first, because sperm would tend to diffuse faster than EGGS, and therefore could not be produced earlier. This gave females the first opportunity to abandon males. Hence the predominance of male parental care in those fish where any care is exhibited.

This idea has been largely discredited because contrary to the hypothesis, externally fertilizing fish typically produce gametes simultaneously or in rapid alternations as the batch of eggs is produced. Thus, the proposed order of gamete release (female first) often fails to match cases of male care. There are also numerous exceptions to the hypothesis even in species where gametes are produced successively by each sex in a clear order. *See also* BEHAVIOURAL ECOLOGY; MATING SYSTEMS. [J.D.R.]

gametophyte The GAMETE-forming, and usually haploid, phase in the LIFE CYCLE of a plant. [G.A.F.H. & I.O.]

Gause's experiments G.F. Gause was an ecologist who worked at the Laboratory of Ecology, in the Zoological Institute at the University of Moscow, in the early 20th century (*see* CHARACTERS IN ECOLOGY). He carried out a series of simple experiments with yeasts and protozoans that are instructive, are of historic interest and set the experimental protocol for many of the so-called 'bottle' experiments. Gause's experimental design was simple. He grew two species, each alone and then together in mixed culture, to determine the course of COMPETITION under laboratory conditions. [B.S.]

gene flow The migration of genetic information occurring between individuals, populations or other spatial units of information (e.g. between two species). [I.O.]

gene pool Two populations are said to share a common gene pool if they are actually or potentially capable of exchanging genes. [I.O.]

general life-history problem Life histories show great variation. Such variation can be encapsulated in the Euler equation, which specifies how the age schedules of survival and reproduction interact to determine a population's or genotype's rate of increase. The general problem is to understand the CONSTRAINTS that determine particular combinations of the two schedules. [D.A.R.]

general-purpose genotype A GENOTYPE, or genome, that confers an ability to exploit a wide range of ecological conditions. The wide ecological and physiological tolerances of general-purpose genotypes may be associated with relatively inferior performance in favourable environments (i.e. the jack of all trades is the master of none). [V.F.]

generalist A species, an individual or (in BEHAVIOURAL ECOLOGY) a STRATEGY that uses a relatively large proportion, or in extreme cases all, of the available resource types. [J.G.S.]

generalized linear models (GLIM or GLM) Models that extend conventional parametric statistical analyses by allowing the user to specify non-normal errors and non-identity link functions. They are especially useful in environmental work for dealing with count data (e.g. log-linear models for CONTINGENCY TABLES), proportion data (e.g. logistic analysis of mortality data), binary response variables (e.g. dead or alive, infected or not, male or female, etc.) and the analysis of survival data (e.g. age at death, using exponential or Weibull errors). The general term used to describe residual variation in GLM is DEVIANCE. [M.J.C.]

generation time The time elapsed from one generation to the next. [H.C.]

genet A genetic individual plant or animal (in contrast to a RAMET); the product of a single zygote (contrast with a MODULE). Plant genets may consist of very large numbers of modules (e.g. tillers of a rhizomatous perennial grass species) and can extend to cover many hectares. The modules may be more or less physiologically independent, depending upon the extent of hormonal and resource transfer between them. [M.J.C.]

genetic bottleneck A substantial decrease in genetic DIVERSITY within a population that is associated with a drastic reduction in POPULATION SIZE. [V.F.]

genetic engineering MANIPULATION of the genome, usually DNA, with various techniques. [P.C.]

genetic fingerprinting A method used to quantify genetic differences among individuals, populations or species. A pattern, or so-called 'fingerprint', is obtained by enzymatically cleaving a protein or nucleic acid and subjecting the digest to two-dimensional chromatography or electrophoresis. *See also* ECOLOGICAL GENETICS; MOLECULAR ECOLOGY; QUANTITATIVE GENETICS. [V.F.]

genetic homeostasis The property of a population to equilibrate its genetic composition and to resist sudden changes. [V.F.]

genetic load A measure of the amount of NATURAL SELECTION associated with a certain amount of genetic VARIABILITY. [I.O. & V.F.]

genetic polymorphism The presence, within a population of a single species, of discrete genetically determined forms or phenotypes such that the occurrence of the rarest of them cannot be due to recurrent MUTATION (a frequency of 1% is usually taken as an arbitrary threshold). [P.M.B.]

genetic revolution One of several terms used in evolutionary studies in connection with the relation of ADAPTATION to species formation. Microevolution by selection of small genetic differences between individuals is a continuous process adapting individuals to local conditions or changing environments. Similar but distinct species differ at many loci, including those which ensure REPRODUCTIVE ISOLATION. Arguably, the transition from one species to another requires some kind of major genomic reorganization—a genetic revolution—in addition to MICROEVOLUTION. [L.M.C.]

genetic systems For a system to evolve by NATURAL SELECTION there needs to be information transmission from one unit of selection to another, and hence 'memory' of that which is favourable. Most commonly this takes the form of a genetic system, which fundamentally involves information transmission in the form of nucleic acids, but may differ in a number of details. A genetic system refers to the organization and method of transmission of the genetic material for a given species. An alternative is the cultural information that is transmitted from one generation of humans to another. Some see this as being divisible into gene analogues—memes—which raises the possibility of having memetic systems. [V.F.]

genetic variance In QUANTITATIVE GENETICS, the genetic variance in a population is that part of the total phenotypic variance of a character that is attributed to genetic causes. The genetic variance, V_G, is the variance of the genotypic values of individuals in the population. [G.D.J.]

genetical ecology The integration of genetics into ECOLOGY. Early ecological research treated all individuals in a POPULATION as identical. In genetical ecology, genetic variation among individuals which can result in variation in PHENOTYPE and in FITNESS or reproductive success is taken into account. It differs from ECOLOGICAL GENETICS, which is primarily the genetics of Mendelian TRAITS in natural populations, in also emphasizing the QUANTITATIVE GENETICS of polygenic traits. [P.M.B.]

genomic imprinting The phenomenon in which genes have differential expression depending on the sex of the parent from which they are inherited: the ALLELES inherited from fathers behave differently from those inherited from mothers, even though they may be genetically identical. [M.Mi.]

genotoxic A broad term that usually describes a CONTAMINANT that has a measurable/observable effect on DNA and/or chromosome structures. [P.C.]

genotype

1 The genetic constitution of organisms.

2 A class of equivalence, employed to classify individuals according to their genetic information. Two individuals are said to have the same genotype if they possess genetic information that is more similar

between them than compared to individuals with a different genotype.
[I.O.]

genotype*environment covariance In QUANTITATIVE GENETICS, genotype*environment covariance is a major experimental problem, especially in field work. The quantitative genetic model of a TRAIT is written as $P = G + E$, i.e. the phenotypic value (P) of an individual derives from genetic (G) and environmental (E) influences. The observed variance over all individuals, the phenotypic variance, V_P, is therefore given by $V_P = V_G + V_E + 2\,\mathrm{cov}(G,E)$, where V_G is the GENETIC VARIANCE (the variation between individuals deriving from their genetic difference), V_E is the environmental variance (the variation between individuals deriving from differences in their environments) and $\mathrm{cov}(G,E)$ is the genotype*environment covariance. [G.D.J.]

genotype*environment interaction In QUANTITATIVE GENETICS, genotype by environment interaction denotes the existence of a difference in the sensitivity of different genotypes to different environments. Differences in the expression of the genotype in different environments leads to different phenotypes over environments, i.e. to PHENOTYPIC PLASTICITY. Genotypes that are phenotypically plastic might differ only in average expression, in which case the REACTION NORMS given by those genotypes are parallel: no genotype*environment interaction exists and the genetic variation is identical in each environment. [G.D.J.]

genotype–phenotype relationship A PHENOTYPE refers to the structural and functional characteristics of an organism determined by its genome, subject to modification by the ENVIRONMENT. Strictly speaking, any part of the organism that is not DNA can be referred to as phenotypic. The selection of genes is mediated by the phenotype, and both genotypic and phenotypic VARIABILITY are necessary for NATURAL SELECTION to act. In order to be favourably selected, a TRAIT must be heritable and the gene(s) responsible for it must augment phenotypic reproductive success as the arithmetic mean effect of its activity in the population in which it is selected. Genes are favourably selected because their phenotypic expression favours the reproduction of their carriers (or close relatives of their carriers). GENOTYPE is commonly used to refer to some part under study of the total genetic complement of an individual, and phenotype as its physical manifestation. In contrast, genome refers to the total genetic complement of an individual, represented by all the chromosomes with their included genes in a somatic cell. [V.F.]

geographic distributions The study of distributions on a regional or global scale, which also requires taking historical factors into consideration in an attempt to understand what has controlled the distributions. Over extensive regions and through geologically considerable periods of time the overwhelmingly most important factors are PLATE TECTONICS, CLIMATE and sea level. [A.H.]

geographic range The extent of the global DISTRIBUTION of a species. *See also* GEOGRAPHIC DISTRIBUTIONS. [P.C.]

geographical barrier A physical barrier to free movement of individuals between populations and hence to GENE FLOW through interbreeding. *See also* SPECIATION. [P.C.]

geographical isolation The separation of a group of individuals or a population from other populations of the same species such that GENE FLOW through interbreeding is impaired. It is caused by GEOGRAPHICAL BARRIERS and can lead to SPECIATION. *See also* ISOLATING MECHANISMS. [P.C.]

geological time-scales The age of the solar system is believed to be 4.7×10^9 years, and the oldest terrestrial rocks are dated at 4.0×10^9 years. Geologists use a hierarchical chronology for the time-span from the origin of fossils to the present day, in which the largest unit is the era. There are only four eras covering the FOSSIL RECORD of life: Cenozoic, $0–65 \times 10^6$ years ago; Mesozoic, $65–245 \times 10^6$ years ago; Palaeozoic, $245–540 \times 10^6$ years ago; and Protozoic, $0.540–2.7 \times 10^9$ years ago. The eras are divided into periods, which are themselves divided into epochs. Table G1 depicts the time-scale spanning the Cenozoic, Mesozoic and Palaeozoic eras, with some of the most important events. *See also* MASS EXTINCTION; TIME. [J.G.]

geometric mean The back-transformed MEAN of logarithmically transformed observations. The formula for calculating the geometric mean (GM) is:

$$\mathrm{GM} = \mathrm{antilog}\left(\frac{\Sigma \log x}{n}\right)$$

where x is each observation in the SAMPLE, n is the number of observations in the sample, and the Greek symbol Σ (capital sigma) is the mathematical notation for 'sum of'. [B.S.]

geometric series

1 Also called a geometric progression, a series of terms in which each term is a constant multiple of the preceding term. The general form of a geometric series is $a, ar, ar^2, ar^3, \ldots, ar^n$, where a is a constant and the sequence has n terms.

2 One of the four commonly used SPECIES-ABUNDANCE MODELS (distributions) (*see* RANK-ABUNDANCE MODELS).

[B.S.]

ghost of competition past A phrase that indicates that the asserted effect of COMPETITION is to eliminate itself by causing NICHE divergence in competing species, and that, as a consequence, it is difficult to prove or disprove that competition is responsible for the divergence. In more general terms, the 'ghost of competition past' could be defined as characteristics of species' niches or distributions, or of COMMUNITY structure that (are believed to) have been

Table G1 The geological time-scale covering the Cenozoic, Mesozoic and Palaeozoic eras.

Era	Period	Epoch	Age (years × 10^6)	Events and noteworthy fossils (dates × 10^6 BP)
Cenozoic	Quaternary	Holocene	0	*Homo sapiens* 0.4, Iron Age 0.003, Atomic Age 0.00005
		Pleistocene	1.6	*Homo erectus* 0.5–1.6, *Homo habilis* 2
	Neogene*	Pliocene	5.2	*Australopithecus* 2.3–4
		Miocene	23	
	Palaeogene*	Oligocene	34	Tree-dwelling primate
		Eocene	53	
		Palaeocene	65	**Mass extinction 65**
Mesozoic	Cretaceous	Late	95	North America, Eurasia, India, Africa, Antarctic/Australia, South America, North America separate
		Early	135	Flowering plants, including herbaceous habit
	Jurassic	Late	152	*Archaeopteryx*
		Middle	180	Coccoliths
		Early	205	**Mass extinction 208**
	Triassic	Late	230	Dinosaurs
		Middle	240	Cycadales
		Early	250	**Mass extinction 245**
Palaeozoic	Permian	Late	260	Reptiles
		Early	300	
	Carboniferous	Silesian	325	Coal formation, winged insects
		Dinantian	355	Amphibia
	Devonian	Late	375	**Mass extinction 367**
		Middle	390	Secondary xylem; tree ferns, 'age of fishes'
		Early	410	Tall vascular plants
	Silurian			Small vascular plants on land
			438	**Mass extinction 439**
	Ordovician			Land plants with spores and cuticles, probable bryophytes
				Corals, first vertebrates
			510	**Mass extinction**
	Cambrian		570	Arthropoda, Echinodermata, Hemichordata, Chordata

* These two units are commonly combined as a single period, the Tertiary.

caused by competition in the past, but are no longer subjected to competition at present. Next to niche divergence, the traces left by competition may include morphological divergence (i.e. CHARACTER DISPLACEMENT *sensu stricto*) and non-random distribution patterns. The ghost becomes even more elusive if we take into account that the failure to find such traces can be explained as a failure to look for the relevant aspects of the community. [J.G.S.]

global stability If a system has an equilibrium to which it will return from any state (other than zero state) then the system is globally stable. [S.N.W.]

global warming CLIMATE warming on a global scale, usually attributed to an enhanced greenhouse effect. There is no doubt that the TEMPERATURE of the

Earth has increased since the end of the 19th century, by 0.3–0.6°C. A trend can be seen in the record from meteorological stations in both the Northern and Southern hemispheres. It is known from historical records that sea levels have risen and that glaciers are receding. There is also no doubt that the atmospheric concentration of carbon dioxide (CO_2) has increased over the last century (it was 280 ppm and is now 360 ppm), and that the concentrations of other GREENHOUSE GASES have been increasing over the past few decades. This observed rise in the concentration of greenhouse gases is believed to be the underlying cause of recent global warming. The link between these two phenomena is explored by mathematical models of the energy balance of the Earth's surface in relation to the optical properties of the ATMOSPHERE. Generally, in such models, any overall increase in the concentration of greenhouse gases causes an increase in the global temperature. Quantitative predictions of the extent of this effect need very elaborate global circulation models (GCMs) that run on supercomputers. Long-term change in the surface air temperatures following a doubling in the CO_2 concentration is predicted to be between 1.5 and 4.5°C, with the 'best' estimate being 2.5°C. There is evidence that vegetation in the Northern hemisphere is already responding to global warming. The spring 'drawdown' of CO_2 caused by photosynthesis is occurring earlier, and so is the spring 'greening' as seen from satellite-borne sensors. *See also* CLIMATE CHANGE. [J.G.]

gnotobiotic Describing a culture, COSM or laboratory animal(s) (e.g. used in toxicity/ecotoxicity tests) in which the exact composition of organisms is known, including, in principle, all associated microbiota. The word derives from the Greek *gnosis*, knowledge, and *bios*, life. [P.C.]

gonochoristic Animals with separate male and female individuals. Plants with this characteristic are dioecious. Thus for animals gonochoristic is the antonym of hermaphroditic. [P.C. & S.C.S.]

goodness-of-fit The extent to which an observed and expected frequency DISTRIBUTION coincide. [B.S.]

gradient The rate of change of a characteristic over distance. [A.J.W.]

gradient analysis The analysis of trends in COMMUNITY composition with respect to environmental gradients in PHYSICAL FACTORS such as pH or moisture. [S.J.H.]

gradients of species richness *See* LATITUDINAL GRADIENTS OF DIVERSITY.

grain The sizes of environmental patches relative to the ecological perception of organisms. Fine-grained ENVIRONMENTS are those in which a single individual encounters all the HETEROGENEITY in a lifetime. A coarse-grained environment is one in which patches are large or long-lasting enough for a single individual to spend its entire life in one patch, followed each generation by DISPERSAL of offspring to the same or other patches. [P.C.]

grasslands Biomes dominated by herbaceous species in which grasses (Poaceae) or sedges (Cyperaceae) are abundant, accompanied by forbs, which account for the highest proportion of plant species found in grasslands, with sometimes scattered shrubs and trees. Grasslands can be classified as natural, semi-natural and artificial. [J.J.H.]

grazing The consumption of the above-ground parts of plants by animals. [G.A.F.H.]

greenhouse gases Gases that contribute to GLOBAL WARMING by absorbing long-wave energy emitted from the Earth's surface—the so-called 'greenhouse effect'. The most widely discussed greenhouse gases are carbon dioxide (CO_2), methane (CH_4), nitrous oxide (N_2O) and the various CFCs, all of which have increased in the last hundred years or so as a result of human activities. Water vapour is also a greenhouse gas, though often overlooked, and so is ozone. These gases differ enormously in their effectiveness as greenhouse gases, and various methods have been devised to express this effectiveness. [J.G.]

Grime's triangle An extension of the *r*–*K* continuum (*see r-* AND *K-*SELECTION) with particular reference to plant species. Grime's triangle is a two-dimensional ORDINATION whose axes are environmental FAVOURABILITY and DISTURBANCE. The model is triangular because one corner of the square ordination, which represents high disturbance and low favourability, is essentially uninhabitable. Grime has recognized three primary plant strategies, which he calls competitors, RUDERALS and STRESS tolerators, and each is characteristic of one of the three apices of the triangle. In favourable conditions where disturbance is low, competitors are favoured. In unfavourable conditions where disturbance is low, stress tolerators are favoured. In favourable conditions with high disturbance, ruderals are favoured. SECONDARY SUCCESSIONS after periodic disturbance exhibit a characteristic trajectory through the triangle; there is a progression from ruderals (short-lived, often EPHEMERAL plants) which are competitively displaced by competitors (tall, rapidly growing plants) and these, in turn, are eventually displaced by stress tolerators (shade-tolerant plants). [M.J.C.]

grooming (preening) Many animals invest a certain proportion of their time in carefully searching and cleaning their skin for ECTOPARASITES and dirt. They either do it directly for themselves (auto-grooming) or do it to someone else (allo-grooming). [C.B.]

gross primary production (GPP) The process of energy, matter or carbon accumulation by the AUTOTROPHS in a community. Its rate, gross primary productivity (P_g or G), equals the sum of net primary productivity and RESPIRATION by the autotrophs. Essentially, gross primary productivity by plants, or other photoautotrophs, represents the rate of total transformation of light energy into chemical energy. In practice, it is estimated for terrestrial plants by correcting net primary productivity with estimates of respiratory losses. [S.P.L.]

group selection The NATURAL SELECTION of some groups of individuals at the expense of other groups. The concept was originally proposed in trying to understand the EVOLUTION of altruistic behaviour. Until quite recently, many social behaviours were interpreted as having evolved 'for the good of the species', with no proper appreciation of the importance of within-species CONFLICTS of interest. In particular, ALLELES producing non-altruistic behaviour may be favoured if their carriers do not pay the costs of altruism, but reap the benefits of the altruistic behaviour of other individuals. Clearly, non-altruists do not act for the good of the species. In an attempt to resolve this paradox, the process of group selection was proposed, according to which groups containing non-altruists are disadvantaged in comparison with groups of altruists. The process was clearly described by V.C. Wynne-Edwards in 1962. Necessary to the evolution of altruism by this process were that migration rates between groups be low, and that groups be short-lived. Most ecologists do not, however, find it plausible that these conditions obtain in the field. This, together with observational support for the rival theory of KIN SELECTION enunciated by W.D. Hamilton in 1964, has led to the general abandonment of group selection as an evolutionary explanation for altruistic behaviour. [R.M.S.]

growth The net increase in size or mass of an organism. [V.F.]

guild A group of species exploiting the same class of RESOURCES in a similar way. [J.G.S.]

gymnosperms The gymnosperms are an ancient group of seed-producing trees whose origins go back at least to the Late Devonian some 400 million years ago. (In contrast the flowering plants, or angiosperms, only appear in the FOSSIL RECORD around 130 million years ago; *see* ANGIOSPERMS.) The distinction between the gymnosperms and angiosperms lies in the reproductive organs and FERTILIZATION processes. In the gymnosperms the seeds are naked (covered only by an often paper-thin integument) and borne on a megasporophyll (the cone) which does not entirely enclose the developing seeds (gymno- derives from the Greek γυμνος, naked). In angiosperms (angio-, αγγειον, vessel) the megasporophyll (the carpel) entirely surrounds the seed. Unlike angiosperms, double fertilization does not occur in gymnosperms thus no true endosperm is formed. The structure of gymnosperm wood is also distinctly different from angiosperms, the gymnosperm water-conducting xylem is composed only of tracheids (though vessels are present in the Gnetales) and the phloem lacks companion cells. [G.A.F.H.]

gynodioecy The condition when some or all populations of a species contain both female (hence the prefix 'gyno-') and HERMAPHRODITE individuals (broad sense) (hence the suffix '-dioecy', referring to the existence of two sexes). [I.O.]

H

habit

1 The regular performance of a particular activity acquired by an animal as a result of learning. In contrast to a conditioned reflex a habit does not require an external stimulus.

2 The characteristic form or mode of growth of an animal or plant.

[P.O.]

habitat The ecological term 'habitat' refers to a place where a species normally lives, often described in terms of physical features such as TOPOGRAPHY and SOIL MOISTURE and by associated dominant forms (e.g. intertidal rock pools or mesquite woodland). Definitions in ecological literature vary widely but there seems to be a consensus for the following:

- 'habitat' entails the description of some key environmental features related to a species or COMMUNITY distribution;
- habitat and VEGETATION classifications may be concordant, but are not always so;
- habitats may have subdivisions as MICROHABITATS, varying in degrees of FAVOURABILITY related to or judged by a species' abundance;
- NICHES within a habitat refer to certain requirements for functional, behavioural and competitive well-being.

[S.K.J. & P.C.]

habitat destruction and alteration, global HABITAT destruction is probably the most important cause of BIODIVERSITY loss. The world's habitats have been so significantly modified by humans that very few can be considered not to bear the mark of human action. Some estimates suggest that more than 70% of the Earth's land surface, other than rock, ice and barren land, is either dominated by human action or at least partially disturbed. Even that which is considered 'undisturbed' is probably subject to some impact.

These anthropogenic impacts take two main forms:

1 conversion of one habitat type to another (e.g. for agriculture or for dwellings);

2 modification of conditions within a habitat (e.g. managing natural FORESTS).

[P.C.]

habitat selection Organisms tend to occur within a restricted range of HABITATS. Although the notion of habitat is open-ended and difficult to define precisely, most organisms are associated with particular biotic and abiotic environments which collectively can be referred to as a habitat. While such associations are apparent, however, it is not always obvious how they arise in practice. [C.J.B.]

habitat structure The physical structure of the ENVIRONMENT that individuals inhabit. [S.J.H.]

hadal Of, or pertaining to the biogeographic region of the deep OCEAN. [S.J.H.]

halophyte A plant of salt-rich habitats such as seawater, SALT MARSHES, MANGROVE swamps and salt DESERTS. [J.R.E.]

handling time The time between perceiving a food item and returning to search for further food—time typically occupied by catching, subduing, handling and sometimes digesting the food item. It is an important parameter, typically symbolized by T_h, in foraging and predator–prey theory (see, for example, HOLLING'S DISC EQUATION). *See also* FEEDING; FOOD SELECTION; OPTIMAL FORAGING THEORY. [H.C.J.G.]

haplo-diploidy A term that generally refers to the GENETIC SYSTEM in which males are normally haploid and females are diploid. More rarely it is used for the alternation of haploid and diploid generations as seen in higher plants. [R.H.C.]

haploidy The condition of having a single set of unpaired chromosomes in each nucleus, as in GAMETES, GAMETOPHYTES and many species of algae, fungi and vascular plants. Can also be a condition of parthenogenetic organisms. *Cf.* DIPLOIDY. [P.C.]

hard selection The term 'hard selection' has changed considerably in meaning since it was coined in the 1960s. Most usage of 'hard' and 'soft' selection refers to multiple-niche models, both one-locus models and QUANTITATIVE GENETICS models. The difference between 'hard' and 'soft' selection has become a difference in formal modelling. In 'hard' selection, genotypic fitness is averaged over environments, and from this overall genotypic fitness allele frequency change for the total environment is computed. Mean fitness over all environments is given by the arithmetic mean over the per environment mean fitnesses, even if within each environment selection would be 'soft'. Ecological considerations show this way to proceed to compute allele frequency change over the total environment and the arithmetic mean fitness to be appropriate for a subdivided population. [G.D.J.]

hard substrata Those substrata that present a surface to which an organism must attach or into which it must bore. [V.F.]

hardening, in plants The process whereby plants become resistant to STRESS by gradual exposure to increasing levels of stress. [J.G.]

Hardy–Weinberg law What is known as the 'Hardy–Weinberg law' is made of two parts, which were discovered independently in 1908 by G.H. Hardy and W. Weinberg. The first part states that if there is no evolutionary force (MUTATION, SELECTION, drift or migration) acting on allele frequencies ('ideal population'), they do not change. It applies to diploid as well as haploid or *n*-ploid species. The second part states that if, moreover, there is random mating, the GENOTYPE frequencies attain equilibrium frequencies in a single generation in a diploid species. [I.O.]

harem A group of two or more females reproductively paired simultaneously with the same male. [J.D.R.]

harm An adverse effect on individuals or an ecological system (populations, communities, ecosystems). [P.C.]

harmonic mean The reciprocal of the MEAN of reciprocals. The formula for calculating the harmonic mean (*H*) is:

$$\frac{1}{H} = \frac{1}{n}\Sigma\frac{1}{x}$$

where x is each observation in the sample, n is the number of observations in the sample, and the Greek symbol Σ (capital sigma) is the mathematical notation for 'sum of'. *Cf.* GEOMETRIC MEAN. [B.S.]

harvest method A technique for measuring PRODUCTIVITY, usually the net primary productivity of herbaceous terrestrial vegetation, for example, grasslands. Sample areas are selected, normally by a randomized block sampling design, and harvested at intervals through the growing season. Harvests are made by clipping the vegetation to ground level. BIOMASS (*B*) in the harvested material is separated from the dead material (*D*), and DRY MASS determined. The technique is most suited to annual crops and wild annuals. During the growing season, net primary productivity (P_n) will be given by the increment in biomass (ΔB) for the time interval (Δt):

$$P_n = \Delta B/\Delta t.$$

[S.P.L.]

harvesting The exploitation of a renewable natural resource, including forestry, fisheries, game ranching and agriculture, although in agriculture a large number of other factors enter into the assessment of the process. [J.B.]

hazard A measure of the potential of a substance or process to cause HARM to human health and the environment. [P.C.]

hazard function A statistical term for the age-specific, instantaneous rate of death in life-table studies. [M.J.C.]

heat balance The balance of heat energy inputs and outputs that determines body TEMPERATURE. Inputs may be from irradiation and from exothermic metabolic reactions within the organism. Outputs are due to radiation, convection, conduction and evaporation. *See also* ENERGY BUDGET. [P.C.]

heathland This term is generally used to denote areas of land dominated by dwarf shrubs, usually of the family Ericaceae. Traditionally, it has been applied to the lowland regions of western Europe in which the heather, *Calluna vulgaris*, is the dominant plant, but its use has extended to related habitats of acidic soil and low vegetation; thus terms such as grass-heath and lichen-heath have evolved. The similar PHYSIOGNOMY, and sometimes shared species, of some Mediterranean vegetation types has led to the application of the term heath to these regions also. [P.D.M.]

heavy-metal pollution Contamination by substances described as heavy metals to an extent that HARM is caused to human health and/or ecological systems. The term 'heavy metal' has been used loosely in this context. [P.C.]

hedgerow A row of shrubs and saplings forming a boundary or barrier; a type of managed natural fencing. Hedgerows act as harbours for natural vegetation and animal life. [P.C.]

hemiparasite

1 In mycology, a fungal hemiparasite (more usually described as a FACULTATIVE parasite) is a fungus which can grow in or on a living HOST from which it obtains its essential energy/NUTRIENTS, whilst having the optional ability to obtain these from non-living sources (and thus it can be cultured on laboratory media).

2 In botany, angiosperm hemiparasites are photosynthetic plants that supplement their nutrition by being parasitic upon another plant; for example the Loranthaceae (mistletoes) and some Scrophulariaceae (e.g. *Euphrasia*).

[J.B.H.]

herbaria Collections of dried, pressed plants, mounted on sheets, labelled and stored for reference. [A.J.W.]

herbicide A BIOCIDE that kills weeds, i.e. unwanted plants. [P.C.]

herbivore Any consumer of living plant materials. *See also* GRAZING; PHYTOPHAGOUS. [P.C.]

heritability A population-dependent parameter, that measures how much of the observed phenotypic variation in a quantitative character is influenced by genetic variation. Heritability, however, says nothing about the mode of inheritance of the QUANTITATIVE TRAIT.

The broad-sense heritability measures what proportion of the total variation is genetic, whether or not this variation is heritable, i.e. whether or not the variations are transmitted to offspring. It is equal to V_G/V_P, where $V_G = V_A + V_D + V_I$ is the total GENETIC VARIANCE, V_D and V_I being the dominance and interaction variances respectively.

The narrow-sense heritability provides an estimate of the similarity of progeny to their parents. It thus allows us to determine the response to selection. It is equal to V_A/V_P, where V_A is the ADDITIVE GENETIC VARIANCE, and V_P is the total phenotypic variance. [I.O.]

hermaphrodite A hermaphrodite animal is one that produces both EGGS and sperm. [I.O.]

heterogeneity

1 The spatial DISTRIBUTION of RESOURCES or some other feature of the ENVIRONMENT of an organism.

2 Another name for species DIVERSITY.

[R.H.S.]

heterosis HYBRID VIGOUR such that an F_1 hybrid falls outside the range of the parents with respect to some character or characters. It is usually applied to size or growth rate. The opposite of heterosis is INBREEDING DEPRESSION. [I.O.]

heterotroph An organism requiring a supply of organic material from its ENVIRONMENT as food. [M.H.]

heterozygosity The heterozygosity, *H*, of an average individual in a population estimates the probability that any individual is heterozygous at any of a specified set of loci. It is calculated as the average frequency of heterozygotes over loci. Heterozygosity differs between populations, species and taxonomic groups. In a number of populations and species, highly heterozygous individuals grow faster. This points to a positive relation between heterozygosity and FITNESS, but not necessarily to heterozygote advantage at each locus *per se*. [G.D.J.]

hierarchy theory Hierarchy theory provides a framework for thinking about and understanding the interactions between processes which operate on different scales. In an ecological context the theory views the spatial and temporal processes in nature as continuous, but with small-scale fast processes at the bottom and large-scale slow processes at the top. It is argued that the relative intensities of interactions between levels can be used to decompose the system into reasonably discrete semi-autonomous parts, or 'holons', which can be viewed within an organizational hierarchy. This hierarchical view places emphasis on the organizational properties which emerge at higher levels in the hierarchy as a consequence of processes operating at other levels. It is argued that, in the same way that the properties of a working clock could not simply be thought of as the sum of its constituent parts, so the EMERGENT PROPERTIES at various levels of ecological organization cannot be derived from studies on the lower-order systems in isolation. *See also* SYSTEMS ECOLOGY. [S.J.H.]

Holarctic realm A biogeographic subdivision comprising the Palaearctic and Nearctic regions, i.e. mid- and high-latitude Eurasia and North America north of tropical Mexico. [A.H.]

holism The doctrine that whole systems cannot fully be understood in terms of the properties of their parts and the laws governing them: for example, populations cannot be understood solely in terms of their constituent individuals, nor communities in terms of their populations. [P.C.]

Holling's disc equation A behavioural model of foraging that assumes the consumer encounters prey at a constant rate and requires a fixed HANDLING TIME to process each prey item. It predicts a type 2 FUNCTIONAL RESPONSE. The equation was originally parameterized for blindfolded human predators SEARCHING for discs of sandpaper that required a fixed handling time; hence the name. *See also* FEEDING; OPTIMAL FORAGING THEORY; PREDATOR–PREY INTERACTIONS. [H.C.J.G.]

home range Individuals of most animal species spend their lives within a circumscribed area which can be thought of as their home range. Empirical determination of home ranges usually involves following an animal about and plotting its movements or points at which it is sighted on a map, then joining the outermost points to form the minimum convex polygon. Of course, such estimates are heavily dependent on how long or often the animal is observed. [C.J.B.]

homeorhesis The maintenance of a pattern of change—as in development—despite environmental disturbances. It is equivalent to HOMEOSTASIS but is concerned with a dynamic rather than a static condition. [P.C.]

homeostasis The tendency of a system (organism, POPULATION, COMMUNITY or ECOSYSTEM) to change its properties in a way that minimizes the impact of outside factors. [P.J.G.]

homeotherm An organism that maintains an approximately constant body TEMPERATURE as environmental temperature varies by the REGULATION of physiological processes, for example shivering and sweating (in ENDOTHERMS), or by appropriate behaviour, for example exposure to solar radiation (mostly in ECTOTHERMS). *Cf.* POIKILOTHERM. [P.O.]

homing The ability of some animals to navigate over long distances to find their way back to a home area (e.g. a nest or breeding site). [R.H.S.]

homology It has been recognized for a long time that most of the TRAITS exhibited by each species also occur in at least one other species. Since the advent of evolutionary theory, two general types of such interspecific similarity have been recognized. Either the species share the traits because they have inherited them from a common ancestor that possessed the trait, or all the species in question acquired the trait independently. When the explanation is common ancestry, we say that the traits are homologous. Thus, a homology is any character that is transmitted genetically from an ancestral species to its descendant species. [D.R.B.]

homozygosity In a random mating population, the probability that an individual is homozygous at any of *n* ALLELES. [G.D.J.]

hormesis Phenomenon in which benefits (e.g. to survival, growth, reproduction) appear to be derived from exposure to small doses/concentrations of substances that are toxic at larger dose/concentrations. [P.C.]

host A living organism that acts as a 'medium' for the growth and development of another organism. [R.C.]

hotshot model An explanation for the evolution of aggregations of males at sites visited by females solely for the purpose of mating (*see* LEK). The model has two main elements: an initial ability of particular males ('hotshots') to attract females, and a tendency for subordinate males to cluster around the hotshot males to mate with females that have been attracted. [J.D.R.]

Huffaker's experiment An experiment that demonstrated the importance of spatial HETEROGENEITY in stabilizing PREDATOR–PREY INTERACTIONS. C.B. Huffaker studied herbivorous mites feeding on oranges; these mites were themselves attacked by predatory mites. With few oranges present, the predatory mites normally drove their prey to extinction. Increasing the number of oranges and making the system more spatially complex (with Vaseline (petroleum jelly) barriers, etc.) led to a persistent interaction. The herbivore was able to escape extinction by colonizing empty patches and reproducing before discovery by the predator. [H.C.J.G.]

humification The microbial DECOMPOSITION of organic matter in soil involves the progressive breakdown of complex molecules into simpler, more soluble ones. Humic acid is a component of these breakdown products and its production is termed humification. [P.D.M.]

'hump-backed' model of diversity A model which describes an observed relationship between PRIMARY PRODUCTIVITY and species DIVERSITY, where highest levels of diversity occur at intermediate productivity levels. [S.J.H.]

humus The natural organic product of DECOMPOSITION of plant material in the SOIL. It is colloidal, composed largely of humic groups—aromatic structures, including polyphenols and polyquinones, which are the products of decomposition, synthesis and polymerization. Humus colloids are negatively charged (especially so at high pH) and usually constitute a large part of the soil's cation exchange capacity. [J.G.]

hunger An abstract motivational response to a physiological need that is commonly restricted to higher animals. It is directly regulated by the animal's requirement for a specific level of nutrition. The intensity of hunger fluctuates temporally and is directly proportional to the activities that are undertaken. [R.C.]

Hutchinson's size-ratio rule According to this rule, coexisting species that have qualitatively similar ecological roles (have to) differ in size by at least a factor of about 1.3 (i.e. if a linear measurement is taken; whereas for masses, the factor should be $1.3^3 \approx 2$). The rule fits in the framework of COMPETITION theory and may be seen as a specification of the principle of COMPETITIVE EXCLUSION, providing a value of the LIMITING SIMILARITY in size between coexisting competitors. [J.G.S.]

hybrid vigour Crosses between inbred lines often lead to hybrid individuals that are larger in size, produce more offspring or yield more seeds, are hardier and stronger. The phenomenon is called hybrid vigour (*see also* HETEROSIS). [G.D.J.]

hybrid zones Narrow zones where genetically differentiated populations meet, mate and produce hybrids. [J.M.Sz.]

hydrogeology The study of water in the ground. It relates to both the saturated zone in the ground, in which all of the void space (porosity) is filled with water, and the unsaturated zone, in which the voids are only partially filled with water. The water is referred to as groundwater. [J.W.L.]

hydrology The study of water, both the quantity and the chemicals and SEDIMENT which it carries, as it moves through the land phase of the hydrological cycle (*see* WATER (HYDROLOGICAL) CYCLE). [J.H.C.G.]

hydrophyte Any plant which is adapted to live in water either continuously or for certain seasons of the year. [R.M.M.C.]

hydroponics The practice of growing plants without soil in which the roots are suspended in water or supported by inert material. [R.M.M.C.]

hydrosere The SUCCESSION from open water via FEN to FOREST or BOG. [P.J.G.]

hydrosphere The part of the Earth covered by liquid or frozen water. *See also* ATMOSPHERE; BIOSPHERE; LITHOSPHERE. [J.H.C.G.]

hydrothermal vent A place on the OCEAN bed, on or near to a midocean ridge, from which water heated by molten rock issues. This is often rich in dissolved sulphides, which are oxidized by chemosynthetic bacteria in the fixation of carbon dioxide (CO_2) and synthesis of organic compounds. Near to the vents there are animal communities that utilize these compounds and may even live symbiotically with the bacteria. [P.C.]

hyperparasitoid A PARASITOID that parasitizes another species of parasitoid. [H.C.J.G.]

hyperspace A multidimensional space which has more than three dimensions so that its geometry cannot be represented by a simple physical model. The hyperspace concept has been employed by ecologists as a means of envisaging data with many attributes (which may be either environmental variables or species). According to this view, a sample can be represented as a point in a hyperspace whose axes correspond to the attributes. [M.O.H.]

hypolimnion The lower layer of deep, cold and relatively undisturbed water in LAKES with THERMAL STRATIFICATION. It often has reduced oxygen, poor light penetration and, therefore, a limited flora and fauna. [J.L.]

I

IC_{50} Statistically derived concentration of a XENOBIOTIC that inhibits some metabolic property (e.g. growth) in 50% of an observed population over a prescribed time in defined conditions. Also expressed in terms of dose, ID_{50}. *See also* EC_{50}; LC_{50}. [P.C.]

ideal free distribution 'Ideal free' theory is concerned with the DISTRIBUTION of competitors across RESOURCES. In its simplest form, it considers a range of HABITATS that vary in the quality of resource (say food or mates) available to a population of competitors. Each competing individual is free to go where it chooses (i.e. is not excluded from particular habitats by dominant or territorial individuals) and 'ideal' in the sense of having perfect information about the quality of each of the habitats (thus the phrase 'ideal free'). The assumption is that competitors will exploit the habitats so as to maximize their reproductive potential. Since the effects of their decisions on reproductive potential will not usually be measurable directly (but see below), it is assumed that the rate of resource acquisition provides an indirect index of reproductive potential. The first competitors to arrive will therefore go to the richest habitat where the resource-acquisition rate will be greatest. As the number of competitors in the best habitat increases, however, the resource begins to deplete and acquisition rate drops. There will therefore come a point at which individuals would do just as well in what was originally the second-richest habitat. The same process occurs in this habitat until it becomes profitable to exploit the third-richest habitat. The upshot is that habitats will eventually be filled in such a way that the rate of resource acquisition by competitors is the same in each. This stable ideal free distribution could be achieved in one of two ways, either as the number of competitors that becomes distributed across habitats in the ratio of initial resource value or through individual competitors visiting all habitats but spending time in each in the ratio of initial resource value. [C.J.B.]

immigration Entry of conspecifics into a population from elsewhere. Leads to GENE FLOW into the population. *Cf.* DISPERSAL; EMIGRATION. [P.C.]

immunity, to parasites The specific RESISTANCE of a HOST to reduce the number or survival of PARASITES. [P.S.H.]

imposex Imposition of male sexual organs on the female. Can occur in marine gastropods as a result of POLLUTION from tributyltin contained in antifouling paints. [P.C.]

imprimitive life history A population with an imprimitive life history does not converge to the STABLE AGE DISTRIBUTION even in a constant environment. This category includes all species with non-overlapping generations, as well as species with OVERLAPPING GENERATIONS that are semelparous and have a fixed juvenile period. [T.J.K.]

imprinting Ethological imprinting is a special learning process during a sensitive phase of an individual's ONTOGENY. [M.Mi.]

***in situ* bioassays** Biological monitors that are either planted out in the environment to be monitored or in close proximity such that continuous or semi-continuous samples can be taken from it. [P.C.]

***in situ* genetic reserves** Contemporary (and much favoured) practices in the conservation of genetic resources include the protection and management of animals and plants in their natural habitats (*in situ* reserves) as opposed to protection in zoos and botanic gardens (*see EX SITU* GENETIC RESERVES), often far removed from the natural habitats. [G.A.F.H.]

inbreeding depression A reduction of FITNESS in a normally outbreeding population as a consequence of increased HOMOZYGOSITY due to inbreeding. Increased homozygosity decreases fitness through the expression of deleterious homozygous recessive TRAITS. [A.R.H.]

incidence functions Incidence functions describe how the PROBABILITY of occurrence of a species varies with selected characteristics of islands, usually their area or SPECIES RICHNESS. They are constructed by plotting the proportion of islands of a given area or species richness that are occupied against their area or species richness. Species incidence generally increases with area, but great variation exists in the shape of incidence functions between species, leading to the (arbitrary) distinction of 'high-S', 'tramp' and 'supertramp' species. [N.M. & S.G.C.]

income breeder An organism that does not use stored body reserves to support reproduction. The alternate mode is a CAPITAL BREEDER. Many small mammals such as shrews are income breeders; because of their high metabolic rate such animals cannot accumulate excess food for STORAGE in their tissues and thus must support lactation by continuous foraging. *See also* LIFE-HISTORY EVOLUTION; TRADE-OFF. [D.A.R.]

increaser A plant that increases in ABUNDANCE in heavily grazed grassland or rangeland, generally as a result of competitor release of unpalatable or GRAZING-tolerant plant species. This competitor release is typically due to the reduction in COVER (or height) of previously superior but more palatable competitors as a result of selective herbivory. Increasers are usually regarded as noxious plants by graziers (they tend to be toxic to livestock, spiny, indigestible or woody). *See also* COMPETITIVE RELEASE. [M.J.C.]

indeterminate growth GROWTH that continues throughout the LIFESPAN of an individual, such that body size and age are correlated. It is usually characterized by a reduction in growth rate past maturation and is typical for many molluscs, many crustaceans, fish, amphibians, turtles and most PERENNIAL plants. [D.E.]

index of biotic integrity (IBI) Attempt to quantify divergences from 'normality' in ECOSYSTEMS (ecosystem harm/disturbances) by measuring a number of attributes of populations and species, in particular: fish and/or macroinvertebrate SPECIES RICHNESS; ratios of native versus non-native species; trophic composition; overall ABUNDANCE and condition. Relatively disturbed sites are compared with IBIs from undisturbed sites in the region. [P.C.]

indicator species This can be used as a general term to indicate species that may be indicative of a COMMUNITY or set of environmental conditions or, very precisely, to refer to species objectively defined as indicators of a phytosociological class, for example sea pink (*Armeria maritima*) for sea cliffs, salad burnet (*Sanguisorba minor*) for chalk grassland and bluebell (*Hyacinthoides non-scriptus*) for oceanic oak woods. The former approach is widely encountered in many areas of biology and is then based on the experience of the person talking or writing or their collective field experience. In the more objective case, the indicator species are derived by some statistical or multivariate technique to specially derive such groupings.

Presence of tolerant species or absence of sensitive species can signal environmental STRESS from natural or human causes. There are two potential problems with this approach.

1 Specificity: organisms that are sensitive/tolerant to one stressor may not respond similarly to others so their presence/absence may not provide an indication of stress.

2 Generality: organisms showing general sensitivity/TOLERANCE do not give sufficiently precise information about environmental conditions.

[F.B.G. & P.C.]

indices of dispersion There are many ways of encapsulating the DISTRIBUTION of individuals in space into a single number (*see* DISPERSION). Most make some assumptions about the distribution and therefore different indices will be appropriate on different occasions. Some of the more commonly encountered indices are described below.

Variance: mean ratio. This is the simplest index of dispersion. If a set of equivalent SAMPLE units of a POPULATION are available (e.g. number of plants in a QUADRAT of fixed size) then calculation is straightforward. The index is easy to interpret as this ratio will be equal to 1 if the population is randomly distributed (i.e. conforms to a POISSON DISTRIBUTION). If the index is greater than 1, then the population is aggregated (i.e. more clumped than random) and if less than 1, the distribution tends towards uniformity (i.e. less clumped than random). Departures from 1 may be tested for significance.

k of the negative binomial. Distributions that are more clumped than random may often be approximated by a NEGATIVE BINOMIAL DISTRIBUTION. In nature, distributions fitting the negative binomial are very common and this index is consequently very useful, although the biological significance of the distribution is still a subject of some debate.

Positive binomial. As for the negative binomial but used in the rare situations where the variance of a set of samples is less than the mean (i.e. individuals are more even than random).

Lloyd's index of mean crowding. This is an index of dispersion that uses data from samples to calculate not the mean number of individuals in each sample but the mean number of individuals encountered by each individual. In other words, it measures the amount of crowding experienced by individuals. If a population follows a Poisson distribution, then MEAN CROWDING will equal the sample mean. If a population is aggregated, then mean crowding will exceed the sample mean. Other indices are based on the concept of mean crowding.

Nearest neighbour analysis. There are several NEAREST NEIGHBOUR TECHNIQUES available. Accurate positional data for every individual in a sample area may be required. The technique usually requires the observer to select individuals at random and then find the distance from that individual to its nearest neighbour. If populations are clumped, then the mean distance from a random set of individuals to their nearest neighbours will be smaller than if the population is randomly distributed. This sort of method is particularly appropriate for SESSILE organisms or analysis of photographic or satellite data. [C.D.]

indirect interaction When the DENSITY of a species alters in a COMMUNITY, the consequent changes in other species result from both direct and indirect interactions. Indirect effects occur when changes in one species affects another through a chain of intermediate species. [S.J.H.]

individual selection A form of NATURAL SELECTION

consisting of non-random differences among GENOTYPES within a population, which leads to changes in the proportions of different genotypes (genes) in the population. *Cf.* GROUP SELECTION. [V.F.]

individualistic hypothesis This proposes that species of plants are distributed individually with respect to biotic and ABIOTIC FACTORS rather than as groups or communities. [P.J.G.]

infanticide Infanticide usually refers to the killing of young conspecifics, including in many cases the infanticidal individual's own offspring. It has been observed in a wide range of species from mammals (including humans), birds and fish to insects, arachnids and crustaceans. In some cases, it is clearly accidental, as in the case of elephant seal (*Mirounga angustirostris*) pups being crushed to death by fighting bulls. In others it appears to be pathological, arising from distortions of parental bonding and care behaviour. However, somewhat counter-intuitively from a functional/evolutionary perspective, infanticide may be an adaptive STRATEGY reaping reproductive benefits for the perpetrator. [C.J.B.]

infochemicals The involvement of chemicals in conveying information in intraspecific and interspecific interactions between organisms is widespread. Examples can be found in interactions among and between vertebrates, invertebrates, plants and microorganisms in the same or different trophic levels. They involve many behavioural and physiological responses, such as SEARCHING for food or oviposition sites, locating a mate, marking a resource area, etc. The chemicals involved are termed 'infochemicals', being defined as chemicals that, in the natural context, convey information in an interaction between two individuals, evoking in the receiver a behavioural or physiological response that is adaptive to either one of the interactants or to both.

Infochemicals can be further classified according to the interaction they mediate. Two criteria are important: (i) whether the interaction mediated by the infochemical is between two individuals from the same species (pheromones) or (ii) between individuals from different species (allelochemicals). In both cases, three subclasses can be distinguished, depending on whether the emitter, the receiver or both benefit from the infochemical (Table I1). These terms are not chemical specific but context specific. One chemical can function as PHEROMONE, kairomone and synomone, depending on the context and the interactants involved. [M.Di.]

information centre hypothesis The hypothesis holds that the exchange of information between individuals at a communal site about the location of rich and EPHEMERAL food patches has been the driving force for the evolution of colonial roosting or breeding behaviour, particularly in birds but also in other taxonomic groups. A foraging individual that is unsuccessful at one occasion could thereby follow a successful individual from the COLONY, and when it is successful it will be followed in turn by unsuccessful ones when leaving the colony. [H.R.]

infracommunity (parasite ecology) The COMMUNITY of PARASITE species that coexist within an individual HOST. [S.J.H.]

inheritance of acquired characters This is the proposition that characters acquired through new use, in the lifetime of an individual, are transmissible from one generation to another. Similarly, characters lost through disuse in one generation do not reappear in future generations. This is known as Lamarckian inheritance, after the French biologist Jean Baptiste Lamarck (1744–1829) (*see* CHARACTERS IN ECOLOGY), who used it as a central principle in his theory of EVOLUTION. Following the rediscovery of Mendel's work, it was increasingly abandoned. A central tenet of evolutionary biology now is that germ lines (genes) are isolated from SOMA (PHENOTYPE) so that the inheritance of ACQUIRED CHARACTERS is not possible; this is known as Weismann's principle (or barrier), after the German biologist August Weismann (1834–1914) who first made it explicit. In fact, towards the end of his life, Charles Darwin (*see* CHARACTERS IN ECOLOGY) also espoused the inheritance of acquired characters because, not having the benefit of Mendel's insights, he presumed that inheritance worked by blending male and female characters, although this would tend to dilute ADAPTATIONS and work against NATURAL SELECTION. Thus he thought the only way that evolution could work was if advantageous characters, acquired during the lifetime of an individual, could be passed on.

However, there were two other features of Lamarckian theory that Charles Darwin would not accept:

Table I1 Definitions of infochemical classes.

Infochemical
A chemical that, in the natural context, conveys information in an interaction between two individuals, evoking in the receiver a behavioural or physiological response

Pheromone
An infochemical that is pertinent to the biology of an organism (1) and that evokes in a receiving organism of the *same* species (organism 2) a behavioural response that is adaptively favourable to:
- organism 1 but not to organism 2: (+,–) pheromone
- organism 2 but not to organism 1: (–,+) pheromone
- organism 1 and 2: (+,+) pheromone

Allelochemical
An infochemical that is pertinent to the biology of an organism (1) and that evokes in a receiving organism of a *different* species (organism 2) a behavioural response that is adaptively favourable to:
- organism 1 but not to organism 2: allomone
- organism 2 but not to organism 1: kairomone
- organism 1 and 2: synomone

1 all organisms were supposed to have innate powers to progress towards a more complex and perfect form;

2 all organisms had an inner disposition that caused the performance of actions sufficient to meet the needs created by a changing environment.

Using these principles, Lamarck explained the evolution of tentacles in the head of a snail as follows: as it crawls along it finds the need to touch objects in front of it and makes an effort to do so; in so doing it sends subtle fluids into extensions of its head; ultimately nerves grow into these and the tentacles become permanent; the acquired tentacles are transmitted from parent to offspring. *Cf.* DARWINISM. [P.C.]

inhibition of succession Occurs where an earlier-invading species reduces the chance (or mean rate) of ESTABLISHMENT of another; *Cf.* FACILITATION. [P.J.G.]

insecticides Chemicals designed to kill insect pests. [K.D.]

integrated conservation strategies Integrated conservation strategies are plans created to conserve species, particularly plants, using all available resources to meet a goal, such as species recovery. For a particular species, integrated strategies might include *in situ* conservation activities (land acquisition, management, habitat restoration), *EX SITU* CONSERVATION (seed or pollen banks, field gene banks or other germplasm storage), research (e.g. reproductive, germination and/or demographic studies) or enhanced legal protection such as through land-use decisions, regulatory listings as protected, or control of plants in commerce. *See also* CONSERVATION BIOLOGY; SEED BANK. [E.O.G. & L.R.M.]

integrated pest management (integrated control) Use of different methods in combination to suppress PEST populations below their economic THRESHOLD. Techniques to be used are selected rationally with an emphasis on their economic and environmental impact. Most successful integrated pest management (IPM) schemes combine chemical and BIOLOGICAL CONTROL with additional methods. [J.S.B.]

intensity of infection The level (number) of macroparasites within an individual HOST (*see also* MEAN INTENSITY OF INFECTION). [G.F.M.]

intention movements Behaviour providing information on the intentions of an individual. [A.P.M.]

interaction variance The interaction variances constitute that part of the GENETIC VARIANCE that arises from interactions between loci in the determination of a QUANTITATIVE TRAIT. [G.D.J.]

interception Precipitation caught by the leaves and stems of a plant CANOPY and evaporated directly back into the ATMOSPHERE without reaching the soil. [J.H.C.G.]

interference A density-dependent reduction in the per capita EFFICIENCY of a PREDATOR or PARASITOID due to direct behavioural interactions. [H.C.J.G.]

interference competition A process of direct or aggressive COMPETITION where individuals interact directly with others in a way that reduces FITNESS, even when RESOURCES necessary for growth and reproduction may not be limited. [R.H.S.]

intermediate disturbance hypothesis Hypothesis that species DIVERSITY in a COMMUNITY is maximized at intermediate levels of DISTURBANCE (where level may refer to frequency, extent or intensity). [P.H.W.]

internal succession A succession that occurs within a COMMUNITY after the creation of gaps by DISTURBANCE, for example in GRASSLANDS on mounds made by gophers or moles, in forests after treefalls, or in Mediterranean-climate scrubland after fire. [P.J.G.]

International Biological Programme (IBP) One of the first attempts to study ecosystems on a world scale through international cooperation. It was initiated by the International Council of Scientific Unions and extended from 1964 to 1974, involving some 40 countries. The sections were: PRODUCTIVITY of terrestrial communities, production processes on land and water, conservation of terrestrial communities, productivity of freshwater communities, productivity of marine communities, human adaptability and the use and management of biological resources. [J.G.]

International Code of Botanical Nomenclature The code is an agreed set of rules and recommendations that govern the scientific naming of plants and fungi. There are also codes for animals and bacteria. The code of nomenclature ensures that the name used by scientists for each organism, or group of organisms, is the same one, so that they can communicate effectively with each other. Organisms are placed in a CLASSIFICATION, a hierarchy of named taxonomic groups (taxa; singular, TAXON). The classification represents relationship. The most important taxon is the species. Every organism belongs to a species with only one correct name, the binomial. The binomial is underlined or italicized and consists of the name of the genus to which the species belongs, written with a capital initial letter, and a specific name, written with a lower case initial letter. Correct names are those that have been effectively and validly published. Valid publication must include a description of the organism and the citation of a nomenclatural type. The 'type' of a species is usually a pressed and dried herbarium specimen. The correct name of a species is the one with priority, the first effectively and validly published name after the publication of Linnaeus' *Species Plantarum* in 1753. Name changes occur as the result of taxonomic research, which discovers better ways to represent relationships in the classification by splitting or lumping together species or genera. [M.I.]

International Code of Zoological Nomenclature (ICZN) These are the regulations governing the scientific naming of animals. They are drawn up and supervised by the International Commission on Zoological Nomenclature (ICZN). [P.C.]

intersexual selection Refers to an individual's ability to attract members of the opposite sex. If the variation in mate quality is similar in both sexes, the sex with the lower potential reproductive rate, usually the female, is the choosy sex. [M.Mi.]

interspecific competition COMPETITION between individuals of different species (heterospecifics). [B.S.]

interstitial The water-filled spaces between the individual particles that comprise aquatic sediments or terrestrial soils form the interstitial habitat, and interstitial organisms are those that can move between the particles without displacing them. [R.S.K.B.]

intertidal zone The shoreward fringe of the seabed between the highest and lowest extent of the TIDES. [V.F.]

intervality If, instead of drawing a line to link predators to prey as in a conventional food-web graph, one links elements that share common prey, the graph obtained is termed a predator, or NICHE, overlap graph. Such graphs can often be collapsed into one-dimensional representations known as interval graphs. FOOD WEBS with predator overlap graphs that can be collapsed in this way are described as interval and this property seems to occur more often in real webs than would be expected by chance. [S.J.H.]

intransitive competition This term is applied to multispecies interactions when a network of INTERSPECIFIC COMPETITION is present. It is used to classify the pattern of interaction present between all directly competing species. For example, species A may displace species B which may in turn displace species C, while species C may also be able to displace species A. Indeterminate unpredictable networks are produced that may promote more diverse and resilient assemblages than if a strict hierarchical structure were present. Tropical reef communities epitomize this type of system, where DISTURBANCE creates an unpredictable spatial environment that prevents the development of fixed competitive hierarchies. Such networks are commonly associated with unpredictable highly heterogeneous environments. *See also* TRANSITIVE COMPETITION. [R.C.]

intrasexual selection COMPETITION among members of the same sex (usually the males) for access to members of the other sex. [M.Mi.]

intraspecific competition Refers to COMPETITION between individuals of the same species. [R.H.S.]

intrinsic mortality Mortality due primarily to internal physiological or metabolic causes such as cancer, heart disease, reproductive stress and degenerative conditions associated with AGEING (SENESCENCE). *Cf.* EXTRINSIC MORTALITY. [T.B.L.K.]

intrinsic rate of increase The POPULATION GROWTH RATE, r, first defined by the demographer A.J. Lotka in 1925 (*see* CHARACTERS IN ECOLOGY). The intrinsic rate of increase is an instantaneous rate and represents the average rate of increase per individual (female) in a population, normally in the absence of crowding effects and shortage of resources; r is also known as the innate capacity for increase and may be calculated from age-specific birth and death rates using the EULER–LOTKA EQUATION or a Leslie matrix. Use of the term 'intrinsic' may be misleading because r depends on both the genetic constitution of a population and the environment in which birth and death rates are expressed, and the intrinsic rate is only achieved when a population has reached its STABLE AGE DISTRIBUTION. Another related measure is population multiplication rate (often specified as λ), which is the factor by which a population multiplies or recedes after a given time-period. (So if $\lambda = 2$, it doubles; if $\lambda = 0.5$, it halves.) *See also* FITNESS; LIFE-HISTORY EVOLUTION; NET RATE OF INCREASE; POPULATION PROJECTION MATRIX. [R.H.S.]

island biogeography The theory of island biogeography was first proposed by Robert MacArthur and Edward Wilson in 1967. This is an equilibrium theory based on the observation that there is a constant relationship between the area of the world's islands and the number of species contained therein. The SPECIES RICHNESS of an island is the result of a balance between the rate of IMMIGRATION of species to the island and the rate of extinction on the island. The resulting equilibrium is dynamic in so far as species are continually arriving and continually becoming extinct, though as time progresses the rates of arrival and extinction diminish. This theory involves a stochastic DYNAMIC EQUILIBRIUM that operates over long time-scales. It also considers island size and distance from the mainland. [A.M.M.]

isolating mechanisms The reproductive characteristics that prevent species from fusing. Isolating mechanisms are particularly important in the BIOLOGICAL SPECIES CONCEPT, in which species of sexual organisms are defined by REPRODUCTIVE ISOLATION, i.e. a lack of gene mixture. Two broad kinds of isolating mechanisms between species are typically distinguished.

1 Pre-mating isolating mechanisms: factors that cause species to mate with their own kind (ASSORTATIVE MATING).

2 Post-mating isolating mechanisms: genomic incompatibility, hybrid inviability or sterility. [J.L.B.M.]

iteroparity Breeding more than once. Iteroparous organisms have a moderate REPRODUCTIVE EFFORT and continue to invest in the maintenance of somatic tissues during and after reproduction, which results in a substantial chance to survive and reproduce again.

There are several types of iteroparous life histories. Iteroparous ANNUALS include most univoltine invertebrates, as well as some annual plants that continue to grow and flower after first seeds have been set. Seasonal breeders live for many years and reproduce repeatedly during specific breeding seasons (e.g. PERENNIAL plants and most vertebrates of seasonal environments). Continuous breeders lack specific reproductive seasons and thus reproduce asynchronously. This category includes species living in aseasonal environments (e.g. TROPICAL RAIN FOREST and deep-sea species, internal parasites), species little affected by seasonality (e.g. humans) and small organisms that complete many generations within one vegetative season (e.g. fruit flies, planktonic crustaceans). For the evolutionary advantages of iteroparity *see* SEMELPARITY. *See also* LIFE-HISTORY EVOLUTION. [T.J.K.]

J

joule (J) SI UNIT of energy, work and heat, defined as the work done when a force of 1 N acts through 1 m. Relation to other units: 1 erg = 10^{-7} J; 1 eV = 1.602×10^{-19} J; 1 calorie = 4.18 J; 1 Btu = 1055 J. [J.G.]

Julian day Julian day is a chronology based on the Julian era with the epoch defined at noon on 1 January 4713 BC. The Julian day is frequently confused with that of DAY NUMBER, which counts the number of days from 1 January within a year. [P.R.V.G.]

K

K-dominance curve A curve that illustrates the DISTRIBUTION of individuals (or BIOMASS) among species in order to graphically represent the species DIVERSITY of a COMMUNITY. Obtained by plotting cumulative per cent ABUNDANCES (or cumulative per cent biomasses) versus species RANK or log species rank. Steepest curves have the lowest diversity. The advantage of *K*-dominance plots is that the distribution of species among individuals can be directly compared with the distribution of species' biomasses, because the units are the same for both plots. *See also* DIVERSITY INDICES. [S.J.H.]

K-selection Selection that operates to maximize POPULATION SIZE. The terms '*r*-selection' and '*K*-selection' are derived from the LOGISTIC EQUATION

$$N(t) = \frac{K}{1 + e^{c-rt}}$$

where $N(t)$ is population size at time t, r is the rate of increase, c is a constant and K is the CARRYING CAPACITY of the environment. Selection that maximizes the rate of increase is termed *r*-selection. It has been hypothesized that there is a TRADE-OFF between selection for TRAITS that maximize r and those that maximize K. However, evidence does not support such a hypothesis. *See also* FITNESS; LIFE-HISTORY EVOLUTION; OPTIMALITY MODELS. [D.A.R.]

k-value Also known as 'killing *k*' or 'killing power', this is a measure of the loss of individuals from a population during a single AGE CLASS or life stage. It is usually followed by a subscripted number that indicates the relevant age or stage. The loss of individuals is usually an actual MORTALITY but may also include a loss due to EMIGRATION. Some authors use a pseudo *k*-value (k_0) to indicate failure to produce the maximum possible number of offspring. A *k*-value is calculated as $\log_{10}$(no. entering a stage) $-\log_{10}$(no. surviving a stage). All *k*-values may be combined to produce k_{total} (or K_{total}) as a measure of generation mortality.

The *k*-values are also used to determine the type of DENSITY DEPENDENCE exhibited by a population. This is achieved by plotting *k*-value against log density on entering a stage. If there is no relationship there is no density dependence, whereas a positive relationship indicates negative density dependence. If the positive relationship has a terminal slope of 1, then the density dependence is exactly compensating (CONTEST COMPETITION); if the slope is less than 1 it is undercompensating; and if greater than 1 it is overcompensating. In the unlikely event that the terminal slope of the line is infinity, this would indicate that pure SCRAMBLE COMPETITION was operating. *See also* KEY-FACTOR ANALYSIS; LIFE TABLE. [C.D.]

Kendall's rank correlation coefficient (τ) A non-parametric statistic that summarizes the degree of CORRELATION between two sets of observations (x and y). [B.S.]

key-factor analysis Key-factor analysis uses the *k*-VALUES calculated from life-table data collected over a number of sampling events (often years) on the same populations to determine which MORTALITY factors have the greatest influence on the variation in the size of a particular population through time. [C.D.]

key organisms A term sometimes used by those involved in environmental management and conservation, which might refer to: importance in dynamics and structure of communities (e.g. as in KEYSTONE SPECIES); important representative of, for example, trophic groups in COMMUNITY; particularly sensitive species in terms of STRESS and/or DISTURBANCE and/or POLLUTION. None of these criteria is exclusive of the others, so that all might be intended. [P.C.]

keystone species A keystone species is one upon which many other species in an ecosystem depend and the loss of which could result in a cascade of local extinctions. The keystone species may supply a vital food resource, as for example in the case of a plant with energy-rich seeds on which many animals rely for food. Or it may be a predatory species that holds in check the populations of certain herbivores which would otherwise overgraze, reduce primary production potential and lead to the loss of other herbivorous animals, together perhaps with certain specific parasites and predators. [P.D.M.]

kin Relatives; i.e. sharing recent common ancestry. *See also* KIN SELECTION. [P.C.]

kin selection (altruism, indirect fitness) W.D. Hamilton pointed out that an allele could enhance its transmission to the next generation either directly, by improving the reproductive success of its own bearer, or indirectly by improving the reproductive success of other individuals carrying a copy of it. The FITNESS of an allele (its rate of spread relative to alternatives in the population GENE POOL) is thus the net result of both these direct and indirect components, thereby constituting what Hamilton termed 'inclusive fitness'.

The importance of inclusive fitness, and its corollary kin selection (the NATURAL SELECTION of alleles through their indirect fitness consequences), is that phenotypic TRAITS that appear to act against an individual's own short-term interests, such as cooperation and 'altruistic' caring and self-sacrificial behaviours, become explicable in terms of genetic selfishness and no longer encourage explanations in terms of GROUP SELECTION (*see* BEHAVIOURAL ECOLOGY) or other ill-founded evolutionary mechanisms. [C.J.B.]

Kolmogorov–Smirnov two-sample test A non-parametric test that examines differences between two distributions, rather than using ranks and comparing medians as in the MANN–WHITNEY *U* TEST. [B.S.]

krill Derived from Norwegian *kril* (young fish) but now used as a common term for euphausids: PELAGIC, marine shrimps distributed globally. Key animal in Antarctic marine ecosystem because of its great abundance and its position in the FOOD WEB between the microalgae upon which it feeds and the large vertebrate predators, whales, seals and penguins, that feed on it. [P.C.]

Kruskal–Wallis test This is the non-parametric equivalent of a one-way ANALYSIS OF VARIANCE. [B.S.]

kurtosis A statistic that measures one type of departure from a NORMAL DISTRIBUTION. Kurtosis is the 'peakedness' of the distribution. A leptokurtic distribution has more observations in the centre and tails, relative to a normal distribution. A platykurtic distribution has less in the centre and tails, relative to a normal distribution. An extreme platykurtic distribution would be a BIMODAL DISTRIBUTION. [B.S.]

k_x Logarithm of the PROBABILITY of survival from one age class to another. *See also* *k*-VALUE. [P.C.]

L

Lack clutch The CLUTCH SIZE that maximizes the number of young fledged assuming that MORTALITY is a function of clutch size, i.e. that the only important interactions are negative density-dependent interactions between siblings within a clutch. If adult survival, survival of offspring subsequent to fledging or the future FECUNDITY of the parents declines with clutch size, then selection will favour a clutch size below the Lack value. [D.A.R.]

lake stratification Layering in water bodies due to different densities of water at different temperatures. May change with season. Extent depends upon physical form of lake, and climate. [R.W.D.]

lakes A hole or basin containing a volume of water that is relatively large compared with its annual inflow and outflow. Lake basins can be small with EPHEMERAL water levels, such as temporary rain-fed rock pools, or extremely large and long-lived, such as Lake Baikal which contains 20% of the world's FRESH WATER and has had a continuous lacustrine history for at least 65 million years. [J.L.]

Lamarckian inheritance *See* INHERITANCE OF ACQUIRED CHARACTERS.

land bridge This is a land connection between continents and between areas within continents. Such connections and their changing temporal distribution have been important factors in the migration routes of terrestrial organisms. Land bridges have therefore influenced patterns of life on Earth today. Moreover, the existence of land bridges in the past was related to the distribution of continental land masses and the movements between them, i.e. CONTINENTAL DRIFT. [A.M.M.]

landscape ecology The study of the interactions between the temporal and spatial aspects of a landscape and its flora, fauna and cultural components. [O.L.G.]

larva A discrete stage in many species, beginning with zygote formation and ending with METAMORPHOSIS. Larvae may develop from planktonic GAMETES, emerge from egg cases or emerge from a parent. [V.F.]

latitudinal gradients of diversity Perhaps the most striking pattern in BIOGEOGRAPHY is the dramatic increase in the DIVERSITY of BIOTA between the poles and the Equator. Although there are conspicuous exceptions to the general rule, for example penguins are much more diverse at high latitudes and most species of salamanders and voles occur in temperate zones, the vast majority of taxa from all the major habitats show the same qualitative trend. This increase in diversity towards the tropics is apparent when either wide geographic regions are considered or when small communities are examined. For example, TROPICAL RAIN FORESTS contain 40–100 species of tree per hectare compared with deciduous forests in the USA (10–30 species) or coniferous forests in northern Canada (1–5 species). Also, many species groups are only found in the tropics, for example New World fruit bats and the Indo-Pacific giant clam. A variety of explanations for this biogeographic pattern have been put forward, none of which is entirely satisfactory when considered in isolation. [S.J.H.]

law of constant final yield Farmers have long known that sowing more seed does not lead to increased YIELDS at harvest. Above a relatively low THRESHOLD seed density, final yield is independent of seeding density; the higher the seed density, the lower the SIZE and FECUNDITY of the adult plants. If density-dependent death of plants occurs when initial plant density is very high, then the yield COMPENSATION may be imperfect (*see* $-^3/_2$ POWER LAW). It is possible that extremely high seed densities cause a reduction in total final seed production (e.g. as a result of outbreaks of fungal pathogens like damping-off disease, *Pythium debaryanum*).

The law of constant final yield has important consequences for plant POPULATION DYNAMICS, because it means that total seed production is independent of the size structure of the breeding population; a few small plants will produce the same seed rain as a large number of small plants. Note that the effect of size distribution on plant MORTALITY is not simple; small plants are much more likely to die than are large plants. [M.J.C.]

layering The process by which a branch of a tree in contact with the ground produces roots at the point of contact and eventually forms a new individual. [G.F.P.]

LC_{50} Statistically derived concentration of XENOBIOTIC that is lethal to 50% of an observed population over a prescribed time in defined conditions. Also referred to as median lethal concentration. Also expressed in terms of dose (LD_{50}). Common indicator of ecotoxicity because when deaths are normally distributed with respect to concentration most of the population will be affected at this concentration and so it can be defined with greatest confidence. [P.C.]

least significant difference (LSD) The smallest difference between any two means that will allow them to be significantly different from each other. Calculated as part of an a posteriori test, after an ANALYSIS OF VARIANCE. [B.S.]

least squares method The method of least squares was discovered independently by two mathematicians, K.F. Gauss and A.-M. Legendre, about 200 years ago. It is used widely as a method for minimizing the residual variation when fitting a relationship or function to a set of observations. [B.S.]

lek A traditional mating ground where adult males defend very small territories devoid of obvious RESOURCES to which females move for the purpose of mating. [J.C.D.]

lekking The MATING SYSTEM or reproductive behaviour of species that mate on traditional mating grounds called leks. Leks are clusters of between five and several hundred small male territories devoid of RESOURCES. They have been documented in about seven species of mammals, 30 species of birds and a number of amphibian, fish and insect species. In lekking populations, sexually receptive females move to these arenas to mate and, once there, appear free to choose between mating partners or territories. Their choice often appears remarkably unanimous, with a few of the males monopolizing the matings. [J.C.D.]

lentic Still water system, i.e. ponds and LAKES. *Cf.* LOTIC [P.C.]

Liebig's law of the minimum The only important factor is the most limiting factor. This is the archetypal single-factor explanation and predicts that the nutrient supplied at the lowest relative rate (i.e. the lowest ratio of supply to demand) is the only nutrient that is rate-limiting. This is generally tested by adding a series of potentially limiting NUTRIENTS, one at a time, and observing whether or not an increase in GROWTH rate occurs. Thus, one might speak of the growth of a given plant as being nitrogen-limited, phosphorus-limited or light-limited in different habitats, but the implication is that a single factor (the 'minimum' factor) limits growth or proliferation in any given environment. [M.J.C.]

life cycle The cycle of developmental events from PROPAGULE through to propagule, for example EGG, LARVA, juvenile, reproductive adult, more eggs, etc. This is sometimes distinguished from life history which is taken to be the sequence of developmental events from one stage in the life cycle to another, for example egg to adult. However, the distinction is often fuzzy, and the two terms are used interchangeably. Another problem encountered in usage is inclusion of a hyphen. The following is most appropriate. When used as a noun, no hyphen is necessary (e.g. the life cycle is semelparous); when used adjectively the hyphen aids clarity (e.g. the life-cycle pattern), indeed in these circumstances some advocate running the words together (e.g. lifecycle pattern). *See also* LIFE-HISTORY EVOLUTION; LIFE-HISTORY TRAITS. [P.C.]

life–dinner principle A metaphor used to explain why prey may always be one step ahead of predators in the evolutionary ARMS RACE of capture and escape. For example, rabbits are running faster than foxes because the rabbit runs for its 'life', while the fox is only after its 'dinner'. If the rabbit fails to escape it will no longer reproduce, while a fox may still achieve reproduction when it loses a race against a rabbit. Therefore, selection is stronger on rabbits than on foxes to improve running ability. The metaphor may apply to any other asymmetric situation. *See also* RED QUEEN'S HYPOTHESIS. [P.S.H.]

life-history evolution The EVOLUTION of those TRAITS that contribute to the major features of the LIFE CYCLE, namely the age schedules of birth and death. More specifically, life-history theory centres upon the evolution of the age and size at first reproduction, the age schedule of REPRODUCTIVE EFFORT and its attendent MORTALITY regime, and the number and size of offspring. Since virtually every aspect of an organism's biology ultimately affects its FITNESS and hence impinges upon the aforementioned traits, there is no precise boundary defining exactly what can be considered within the purvue of life-history theory.

A central feature of the theory of life-history evolution is the hypothesis that variation is circumscribed by the existence of trade-offs. For example, increasing the size of offspring can only be done at the expense of the number of offspring. Given the existence of trade-offs, evolution will lead eventually to that combination which maximizes fitness. In order to avoid the problem of tautology, the measure of fitness must be defined a priori. The most frequently used measures are the INTRINSIC RATE OF INCREASE, r, or the net reproductive rate, R_0. The former defines the rate of population increase in the absence of density-dependent factors, while the latter refers to the specific situation of a stable population. The values of these parameters are obtained by solving the EULER–LOTKA EQUATION. In both cases density-dependent factors are assumed not to influence the trait under study. This does not necessarily mean that the population is not under density-dependent control. Suppose, for example, that in some organism POPULATION SIZE is determined by density-dependent processes that occur in the larval stage but the characters under selection at this stage are genetically uncorrelated with adult traits. In this circumstance analysis of the evolution of adult traits can proceed using the net reproductive rate, but analysis of larval traits must take into account the effects of density. Given a measure of fitness and a set of trade-offs we can predict which combination of traits will maximize fitness.

An alternative approach to the assumption that some specific fitness measure is maximized is to use a model incorporating the mode of transmission of the traits from generation to generation (e.g. single locus or quantitative genetic models) plus a criterion for the relative contribution of different genotypes to the next generation (i.e. a SELECTION COEFFICIENT). Both approaches have strengths and weaknesses but, in general, both lead to the same predicted evolutionary equilibrium. *See also* AGEING; EVOLUTIONARY OPTIMIZATION; OPTIMALITY MODELS; REPRODUCTIVE VALUE. [D.A.R.]

life-history traits These are major features of an organism's LIFE CYCLE, most directly related to the schedule of birth and death rates and thus to Darwinian FITNESS (often referred to as life-history or life-cycle strategies). This somewhat loosely defined set includes body size, patterns of GROWTH and STORAGE, age at maturity and METAMORPHOSIS, rate of SENESCENCE, LIFESPAN, REPRODUCTIVE EFFORT, mating success, number, size and sex ratio of offspring, PARENTAL INVESTMENT. [T.J.K.]

life table A tabulation of the patterns of MORTALITY, and thus also of survivorship, of the individuals in a POPULATION, equivalent to an actuarial table used by life insurance companies to determine the life expectancy of an applicant. There are two distinct forms, the COHORT LIFE TABLE and the STATIC LIFE TABLE. Both contain measures of the POPULATION SIZE and of mortality with respect to stage, age or AGE CLASS (see Table L1 for definition of terms usually used in life tables) and frequently incorporate fertility schedules or FECUNDITY SCHEDULES. Life tables embody, together with RECRUITMENT curves and survivorship curves, the data necessary for demographic analysis (e.g. calculation of BASIC REPRODUCTIVE RATE R_0, k-factor analysis). *See also* a_x; DEMOGRAPHY; d_x; e_x; l_x; q_x. [A.J.D.]

lifespan Measure of the duration of life, used variously to describe either the maximum or average duration of life in a population or the duration of life of an individual organism. *See also* AGEING; DEMOGRAPHY. [T.B.L.K.]

lifetime reproductive success The total number of offspring produced by an individual over its lifetime, often taken as a measure of FITNESS. This is exact only in the case of average lifetime reproductive success of a GENOTYPE in a population that is constant in size. [G.D.J.]

limiting similarity The smallest difference between competitors that allows their COEXISTENCE. [J.G.S.]

limnology Derived from the Greek *limnos* meaning SWAMP, lake or MARSH, the word 'limnology' was used originally to define the study of LAKES. Limnology was later expanded to include all inland waters (fresh and salt) contained within continental boundaries including standing (LENTIC) ecosystems such as lakes and ponds and running (LOTIC) waters (springs, streams, creeks, RIVERS). Brackish waters in river estuaries are also included in limnology as are EPHEMERAL (temporary) water bodies such as pools, tree holes, etc. [R.W.D.]

Table L1 Definition of terms usually used in life tables.

x	Age (or stage) of individuals
a_x	Number of individuals of age (or stage) x
d_x	Proportion of individuals dying at age (or stage) x
e_x	Life expectancy of individuals at age (or stage) x
l_x	Proportion of the organisms at age 0 surviving to age x
q_x	Mortality rate at age (or stage) x

Lincoln index A basic equation used to estimate numbers in a mobile POPULATION in which marked animals are released and recaptured. Named after F.C. Lincoln who developed it in relation to the American bird banding programme established in the 1930s. *See also* CAPTURE–RECAPTURE TECHNIQUES; SAMPLING METHODOLOGY/DEVICES. [L.M.C.]

Lindeman efficiency Lindeman efficiencies are the efficiencies with which energy is transferred between trophic levels (also known as ecological or transfer efficiencies) and are the fundamental efficiencies that set the availability of energy to the next TROPHIC LEVEL. [M.H.]

lineage Cohesive groups of organisms, whose cohesion can be based on common history, developmental CONSTRAINTS, STABILIZING SELECTION and/or interbreeding. The most inclusive evolutionary lineages are species. *See also* CLASSIFICATION; SPECIATION. [D.R.B.]

linear models Models in which the MEAN response is a linear combination of the PARAMETERS and which follow the general form:

$$\begin{aligned} y = \alpha &+ \beta_1(\text{something involving just } x \text{ values}) \\ &+ \beta_2(\text{second quantity involving just } x \text{ values}) \\ &+ \cdots + \beta_k(k\text{th quantity involving just } x \text{ values}) + \varepsilon \end{aligned}$$

Linear models do not have inherent scaling in that any solution for x and y can be scaled up or down and still be a solution. Linear differential models have the advantage that they can be solved analytically. *See also* REGRESSION ANALYSIS. [V.F.]

Linnaean classification *See* BINOMIAL CLASSIFICATION.

lithosphere This is simply defined as the Earth's CRUST. [E.A.F.]

littoral zone

1 For LAKES, the edge, sometimes defined imprecisely as the extent to which SAMPLING is possible before water overflows the boots, and sometimes more precisely as the area in which PHOTOSYNTHESIS is possible by benthic primary producers. *See also* WITHIN-LAKE HABITAT CLASSIFICATION.

2 For the seas, also called the eulittoral or INTERTIDAL ZONE. The limits of this zone are sometimes considered to lie between mean high-water and mean low-water levels. Alternatively, the zone may be considered to extend from the level of the highest spring high tide to the lowest spring low tide. Additionally, the supralittoral zone (splash or spray zone) may be used to refer to those levels on the shore only submerged during spring high TIDES. [V.F.]

living fossils Species of animals and plants that have changed little over long periods of evolutionary history to the present. Examples are the lungfish, alligators, snapping turtles and the aardvark. [P.C.]

local stability A system is locally stable if it returns to an equilibrium when subjected to a very small perturbation away from that equilibrium. [S.N.W.]

log normal distribution One of the four commonly used SPECIES-ABUNDANCE MODELS (distributions). [B.S.]

log series distribution One of the four commonly used SPECIES-ABUNDANCE MODELS (distributions). [B.S.]

logistic equation Simplest equation relating POPULATION GROWTH RATE to POPULATION DENSITY. If N represents population density and t is time, then $1/N\ dN/dt$ represents population growth rate and this is assumed to be a straight-line function of N. This is the logistic equation:

$$\frac{1}{N}\frac{dN}{dt} = r - \frac{r}{K}N$$

or:

$$\frac{dN}{dt} = rN\left(1 - \frac{N}{K}\right)$$

where r is a constant representing population growth rate at low population density and K is the equilibrium population density, which is referred to as the CARRYING CAPACITY of the environment in which the population lives. The resulting sigmoidal curve of population density against time (Fig. L1) is referred to as the logistic curve. Without DENSITY DEPENDENCE the population would grow exponentially, $N = N_0e^{rt}$, where N_0 is population density at time zero (Fig. L1). The logistic equation is sometimes used in a discrete-time form, generally written as:

$$N_{t+1} = \frac{N_tR}{1 + aN_t}$$

where $R = e^r$ and $a = (R - 1)/K$.

This simplistic treatment assumes the population has no AGE STRUCTURE, so that all individuals are equally capable of reproducing at any time. *See also* LOGISTIC MODEL; POPULATION DYNAMICS; POPULATION REGULATION. [R.M.S.]

logistic model The simplest model of density-dependent population growth, in which the POPULATION GROWTH RATE decreases in a straight-line relationship with increasing POPULATION SIZE. The two parameters of the logistic model are the INTRINSIC RATE OF INCREASE r and the equilibrium population level or CARRYING CAPACITY K. The continuous-time (differential equation) version of the logistic model shows SIGMOID GROWTH with a smooth approach to the carrying capacity. The discrete-time (difference equation) version in contrast shows a range of dynamic behaviour, from STABLE EQUILIBRIUM through regular cycles to CHAOS, as a consequence of DELAYED DENSITY DEPENDENCE. *See also* LOGISTIC EQUATION; LOTKA–VOLTERRA MODEL. [R.H.S.]

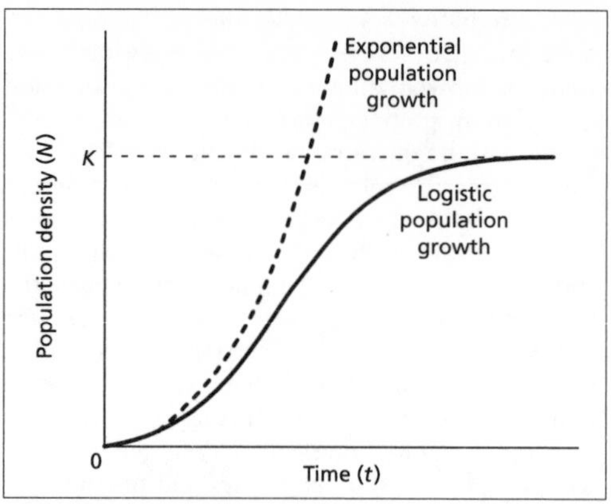

Fig. L1 The two simplest types of population growth, with (logistic) or without (exponential) density dependence. K is the carrying capacity of the environment.

Long-term Ecological Research Program (LTER) Programme in the USA that was initiated in the 1970s by the National Science Foundation. It focused on examining ecological processes and problems with temporal or spatial scales that could not be addressed by more traditional research projects with shorter time horizons. It now involves 18 sites that range from the Arctic/Antarctica to temperate and tropical. [P.C.]

lotic Running waters, such as RIVERS and streams. *See also* LIMNOLOGY. [J.L.]

Lotka–Volterra model Vito Volterra was an Italian mathematician and Alfred Lotka was an American mathematician who published in the 1920s and 1930s. Their names are associated with two basic models that have had a major impact on how ecologists think about INTERSPECIFIC COMPETITION and predation. (*See also* CHARACTERS IN ECOLOGY.) The competition model is based on the LOGISTIC EQUATION and introduced competition coefficients (α)—factors to allow one species to be represented in numerical equivalents of the other species. The predator–prey model was based upon the exponential model of population dynamics and illustrates, amongst other things, how predator–prey dynamics can show oscillations. [B.S.]

lottery model A model of COMPETITION for primary space, originally proposed to explain the COEXIS-

TENCE of the large numbers of territorial coral reef fish species. The model assumes that space is limiting and that there are many more larvae trying to settle and establish territories than there are spaces. Those larvae that do manage to find suitable territories are assumed to hold them for the rest of their lives, so a territory only becomes available when an individual dies. Larvae are highly dispersive, so that the number of a given species seeking territories in any one area is not thought to be related to the number of resident adults and newly available territories are assumed to be allocated to the first larva to arrive. Analyses show that the lottery model can account for coexistence on a local scale and, if it applies in nature, RESOURCE PARTITIONING between species ought to be weak or absent. [S.J.H.]

lunar cycles Patterns of movement, FEEDING and reproduction of marine organisms of the seashore are often tightly linked to the tidal regime and hence to lunar cycles. Many shallow-water tropical reef fish have lunar-periodic reproduction.

The mechanisms by which organisms synchronize their activities to lunar cycles are still under investigation. Several physiological clocks (circadian and those with long-term rhythms) seem to be involved that are set by receptors for light or for water disturbances. [W.W.W.]

l_x In a LIFE TABLE, the proportion of the number of individuals in AGE CLASS or stage zero surviving to the start of age class or stage x. Calculated as a_x/a_0. [A.J.D.]

lysimeter Experimental system consisting of a block of soil to which an aqueous liquid of known quality is added and from which the routes and rates of loss, by percolation, evapotranspiration and RUN-OFF, can be budgeted. The quality of percolate is often tested ecotoxicologically. Amongst other things, it may be used to test the potential of chemicals sprayed on land and landfill materials to contaminate and pollute ground and surface waters. [P.C.]

M

MacArthur–Wilson model of island biogeography A model proposed to explain the numbers of species observed on islands based on the idea that there is a dynamic balance between rates of IMMIGRATION and extinction. [S.J.H.]

macrocosm Large, isolated and controlled multispecies system, often outdoors, used for EXPERIMENTAL ECOLOGY and ecotoxicity testing. Some attempt has been made at being more precise in specifying size of system for which the term applies, but the separation of macrocosm from MESOCOSM is not sharp and probably not really helpful. [P.C.]

macroevolution The EVOLUTION of great phenotypic changes, usually great enough to allocate the changed LINEAGE and its descendants to a distinct genus or higher TAXON. Much of the modern synthesis of the 1930s and 1940s was devoted to demonstrating that the mechanisms of MICROEVOLUTION (i.e. genetic changes within populations and species) can adequately account for macroevolutionary phenomena. More recent debate surrounding the PUNCTUATED EQUILIBRIUM model of evolution has renewed interest in the relationship between microevolutionary and macroevolutionary processes. [V.F.]

macrofauna

1 Generally, fauna in SAMPLES that are visible unaided by lens.

2 More specifically, the extraction of animals from marine sediments almost invariably requires the use of sieves and hence the mesh size of the sieve used determines the SIZE of animal retained—the macrofauna are defined as those benthic animals retained by a 0.5-mm mesh and, by extension, the same term is often used for animals of the same size from other benthic habitats (*cf.* MEIOFAUNA; MICROFAUNA). [R.S.K.B.]

macronutrients These are the elements (other than carbon, hydrogen and oxygen) required by living organisms in substantial amounts: nitrogen, phosphorus, potassium, calcium, magnesium, sulphur and chlorine (*cf.* MICRONUTRIENTS). [K.A.S.]

Malthusian parameter The rate of increase of a population growing at its STABLE AGE DISTRIBUTION with constant age-specific survivorship (l_x) and reproduction (m_x). The Malthusian parameter is equivalent to the INTRINSIC RATE OF INCREASE, *r*. [H.C.]

mangrove The word can be applied either to the plant belonging to a group of trees that dominate tropical intertidal forests or to the forest habitat itself. Some ecologists prefer to use the term 'mangal' specifically for the habitat in which mangrove plants grow rather than for the plants themselves. [P.D.M.]

manipulation Manipulation in an evolutionary context is a metaphor for the effects of one entity on the behaviour (or other phenotypic attribute) of another that cause the second entity to respond in the apparent interests of the first. Manipulation can occur at a number of different levels, hence the hedging term 'entity'. Where it occurs at the level of the individual it is an example of what Richard Dawkins terms 'extended phenotypic' effects: adaptive phenotypic consequences of ALLELES outside the physical boundary of their immediate bearer. Manipulation in this sense occurs in many interspecific and intraspecific interactions, from host–parasite and predator–prey associations to relationships between parents and offspring and mated pairs. [C.J.B.]

Mann–Whitney *U* test (Wilcoxon rank sum test) This is the non-parametric equivalent of the *z(d)* TEST or *t* TEST. However, it is used to compare two medians, rather than two means, and is suitable for UNMATCHED OBSERVATIONS. [B.S.]

marginal value theorem (MVT) The MVT predicts how long an animal should remain exploiting a specific source of a resource before leaving to exploit another. Applicable to any patchily distributed RESOURCES such as food, hosts or mates, the MVT is best known in the context of OPTIMAL FORAGING THEORY, where it is concerned with the exploitation of food. Resources are assumed to occur in patches separated by unproductive habitat. The patches are encountered randomly but are not revisited and are depletable within a single visit by the exploiter. Optimization is based on the ratio of gain (resource acquisition) to cost (time spent travelling to the patch and exploiting the patch itself). The optimal exploitation (patch-residence time) maximizes the instantaneous rate of resource acquisition (marginal value). This occurs when, because of depletion (or any other cause of diminishing returns), the instantaneous gain rate falls to the average for all patches within the habitat. [R.N.H. & J.V.A.]

mariculture The artificial cultivation of marine species. [V.F.]

marine snow Amorphous particulate matter derived from living organisms. It is formed from mucus released by large planktonic filter feeders, which produces flakes. The flakes become colonized by bacteria and their flagellate and ciliate grazers. Sinking of this material, especially in dense concentration, looks like a snowfall. [V.F.]

marking, territorial The marking of the boundaries of an exclusive area (a territory) by visual, vocal, olfactory or other signals. [A.P.M.]

marsh This term causes confusion in wetland nomenclature because it is used in very different senses in different parts of the world. In North America, the expression is equivalent to the European term 'SWAMP'. In Europe, 'marsh' is used to refer to ecosystems with periodically waterlogged mineral soils but with little or no PEAT formation. The soils often take the form of groundwater gleys and the vegetation, which is usually herbaceous, is characteristically capable of tolerating waterlogging. [P.D.M.]

mass balance Accumulation of total mass in a system is equal to the difference between input and output. May also apply to a particular substance, and will apply to elements. Based on the principle of conservation of mass. [P.C.]

mass extinction Mass extinctions are substantial BIODIVERSITY losses that are global in extent, taxonomically broad and rapid relative to the average persistence times of species in the taxa involved. *See* EXTINCTION CRISIS. [M.A.B.]

matched observations Pairs of observations made on the same experimental or observed unit. For example, if we measured egg production (per female) in a group of fruit flies on one day (under one set of conditions) and then measured daily egg production in the same group of females on another day (under perhaps another set of conditions), the data on egg production would be paired because two measurements had been made on each female. *Cf.* UNMATCHED OBSERVATIONS. [B.S.]

mate-defence polygyny A MATING SYSTEM where individuals compete for members of the opposite sex without using fixed territories. Typically, this occurs where the male provides little or no PARENTAL CARE, and breeding resources are not sufficiently clumped in space or time to select for male TERRITORIALITY. Instead, males may move with groups of females in home ranges that overlap with those of other males. [J.D.R.]

mate-guarding The close following of females by their mates during the female's period of fertility. The function of mate-guarding is to protect the PATERNITY of the guarding male. Mate-guarding occurs in a variety of organisms including molluscs, crustaceans, insects, fish, amphibians, reptiles, birds, mammals and humans. [B.B.]

mate recognition Organisms that form a pair bond during the reproductive period or even for life, such as many birds, have the ability to recognize their partners as individuals. This is mate recognition and should be differentiated from SPECIFIC-MATE RECOGNITION, where an appropriate mate is recognized. [H.E.H.P.]

maternal effect The phenotypic value of an individual for a certain trait might be influenced by its mother's GENOTYPE or PHENOTYPE. If so, the QUANTITATIVE TRAIT shows a maternal effect, rather like a phenotypic time-lag. [G.D.J.]

mating systems The numbers of mates obtained by males and females and the nature of reproductive relationships between them in species or populations of animals. Males may practise MONOGAMY (one mate, e.g. gibbons) or POLYGYNY (many mates, e.g. sage grouse); females too may be monogamous (one mate) or engage in POLYANDRY (many mates, e.g. marmosets and tamarins). Bonds between males and females may be lifelong (e.g. swans), seasonal (e.g. mallards) or non-existent (e.g. fallow deer). [J.C.D.]

matrix models These are potentially a very powerful tool for the investigation of POPULATION DYNAMICS that take into account the age or stage structure of the population in question. The earliest uses of matrix algebra in the study of population dynamics come from the use of life tables, where the survival of different ages or stages in the life history of an organism are calculated separately. [C.D.]

matter flux Matter flux is one of two fundamental processes that occur in the BIOSPHERE, the other being energy flux. Alternative descriptions for matter fluxes include BIOGEOCHEMICAL CYCLES and nutrient transfers. [A.M.M.]

maximum likelihood methods Maximum likelihood is the most general method of making a statistical estimate; it is an alternative to the LEAST SQUARES METHOD. Given a model (M) and some data (D), the likelihood of a given value of an estimated parameter (V) is the PROBABILITY of the data given the model and that value of the parameter, $P(D: V,M)$. The probability of the data is thus considered as a function of the parameter. The probability of all possible datasets must add up to 1, but when the data are held constant and the parameter is varied the different values of $P(D: V,M)$ do not have to add to 1; they are called likelihoods to distinguish them from probabilities. The maximum likelihood method chooses that value V that maximizes the probability that the observed data would have occurred. [S.C.S.]

maximum sustainable yield This key concept is largely theoretical but has a sensible underlying idea. A population's ability to sustain a harvest is limited: that limit is the maximum sustainable yield (MSY). The key to this process is DENSITY DEPENDENCE, i.e. as a population is reduced its demographic performance improves and the surplus accrued by that improved performance (more births, less deaths, more growth) can be used to generate a SUSTAINABLE YIELD. [J.B.]

mean This is the most common MEASURE OF CENTRAL TENDENCY or average, familiar to most people. It is calculated by summing all the individual observations in a SAMPLE and dividing this sum by the number of observations in the sample. The formula for calculating the mean ($\bar{x}$ or x-bar) is therefore:

$$\bar{x} = \frac{\Sigma x}{n}$$

where x is each observation in the sample, n is the number of observations in the sample and the Greek symbol Σ (capital sigma) is the mathematical notation for 'sum of'. [B.S.]

mean crowding Mean crowding (m^*) is the average DENSITY experienced by a randomly chosen individual in a population. [B.S.]

mean intensity of infection Average number of macroparasites per HOST in a defined host population (*see* INTENSITY OF INFECTION). Macroparasites typically have a skewed DISTRIBUTION between hosts, such that most PARASITES are found in a small proportion of the host population. [G.F.M.]

measure of central tendency (average) A single number that is representative of all the numbers in a set of observations. Three types of average are commonly used: the MEAN, the median and the MODE. When the observations follow a symmetrical DISTRIBUTION then the mean, median and mode have the same value. In a non-symmetrical (skewed) distribution the mean is moved towards the extended tail of the distribution. [B.S.]

Mediterranean scrub/forest Mediterranean-type climates are typified by hot dry summers and mild moist winters. This type of climate is found in five different parts of the world: the Mediterranean Basin, South Africa, southern Australia, California and Chile. The vegetation is characterized by dwarf shrubs, scrub or woodland. [P.D.M.]

medium number systems, problems of prediction Biological communities are often termed 'medium number systems' because of the number of species they contain. The reason for the distinction lies primarily in the mathematical approaches that can be used to describe species dynamics. For large number systems describing, for example, the behaviour of molecules in a chemical reaction, average or probabilistic approaches can be used. In contrast, when there are very few interacting elements, models can be constructed using a set of equations that specifies the behaviour of each individual element. For medium number systems, it has been argued that neither of these approaches can be used: there are too many elements (species) to construct and solve equations using the small number approach and too few for probabilistic behaviour to provide accurate predictions on the timescales of interest. The absence of a satisfactory framework for modelling medium number systems can be a principal limitation for quantitatively predicting the responses of species populations in a COMMUNITY context for some classes of problem. [S.J.H.]

megaplankton ZOOPLANKTON of between 20 and 200 cm in size. All are metazoans. [V.F.]

meiofauna The extraction of animals from marine sediments almost invariably requires the use of sieves and hence the mesh size of the sieve adopted determines the SIZE of animal retained. The meiofauna are defined as those multicellular benthic animals that pass through the 0.5-mm mesh used to collect the MACROFAUNA and, by extension, the same term is often used for animals of the same size from other benthic marine habitats, although it is not used in respect of the PLANKTON or terrestrially. [R.S.K.B.]

melanism Many animals, especially species of moths and other insects, exhibit melanic forms that are black or dark in colour due to the deposition of melanin pigment in the cuticle or skin. *See also* ECOLOGICAL GENETICS. [P.M.B.]

Mendelian inheritance The laws of Mendel concerning inheritance are the law of segregation (first law) and the law of independent assortment (second law).

The first law concerns the particulate nature of the genes. Genes inherited from the parents keep their separate identity when occurring together in one diploid individual. Alternate versions of the genes, called ALLELES, account for variation in inherited characters.

The second law concerns crosses of two or more characters. If two characters, like seed colour and seed shape in peas, are inherited independently, the ratios of the combination of characters can be found by combining the ratios from the monohybrid crosses. [G.D.J.]

Menge–Sutherland model A theoretical model of community REGULATION that attempts to describe how the relative importance of structuring process (i.e. COMPETITION, predation and DISTURBANCE) alter along gradients of environmental STRESS and for different trophic levels within the community. A later elaboration of the model includes the effects of differing levels of RECRUITMENT. [S.J.H.]

mesocosm As MACROCOSM, but 'medium sized' and often indoors. [P.C.]

mesopelagic The part of the OCEAN water column extending from the bottom of the epipelagic zone (*c.* 200–300 m) to about 1000 m depth. It corresponds approximately with the dysphotic (twilight) zone and is located seaward of the shelf-slope break. [V.F.]

mesoplankton ZOOPLANKTON species that are between 0.2 and 20 mm in size. [V.F.]

metamorphosis Transitional stage in organisms with a COMPLEX LIFE CYCLE, involving the remodelling of a larval form into an adult morphology. [P.A.]

meteorology Originally the science of things aloft (from the Greek), it has come to mean the science of the physical nature of the lower ATMOSPHERE, especially the behaviour and forecasting of weather systems. [J.F.R.M.]

microbial ecology The science that specifically examines the relationships between microorganisms and their biotic and abiotic environments. [V.F.]

microbial loop This refers to the regeneration of NUTRIENTS and their return to the FOOD CHAIN that is mediated by bacteria and protozoans. Bacteria can utilize both particulate DETRITUS (FAECAL PELLETS, dead organisms) and DISSOLVED ORGANIC MATTER (excretory products and microbial exudates). The bacteria are in turn fed upon by protozoans, which are fed upon by higher trophic levels (e.g. copepods). Of the total PHYTOPLANKTON production converted to dissolved organic matter, 10–50% is consumed by bacterioplankton. Much of this material is cycled through the microbial loop. [V.F.]

microclimate The CLIMATE in small places, usually considered over a spatial scale of a few centimetres or metres and over short periods of time (hours). [J.G.]

microcosm As COSM, but 'small' and usually indoors, in laboratories. [P.C.]

microevolution EVOLUTION of small changes within populations, given sufficient time, can lead to considerable divergence between them and ancestral groups and possibly once closely related but now isolated populations. Changes occur by NATURAL SELECTION, and microevolution is often considered to be synonymous with (neo) Darwinian evolution. *Cf.* MACROEVOLUTION. [P.C.]

microevolutionary trade-off Refers to a special case of correlated evolution for which there is a balance between the perceived costs and benefits derived from the evolution of a given trait. One class of examples comes from functional morphology. For instance, in the evolution of the limbs of tetrapod vertebrates there is a necessary trade-off between the potential output velocity of the limb, which might be adaptive for running, and the potential output force of the limb, which might be adaptive for jumping or digging. This trade-off is a consequence of the contributions of the relative lengths of the 'in' and 'out' arms of the limb, when the limb is modelled as a lever system. A relatively long 'in' arm can increase the output force of the limb, but it necessarily decreases the potential output velocity. The morphology of an organism thus represents a compromise associated with the importance of these different aspects of performance in its lifestyle. A 'specialist', such as a cheetah (specialized for running) or an armadillo (specialized for digging), is an organism for which one aspect of performance has had an overriding influence on FITNESS and the evolution of life-history patterns. [D.N.R.]

microfauna The extraction of animals from marine sediments almost invariably requires the use of sieves and hence the mesh size of the sieve used determines the SIZE of organism retained. The microfauna are defined as those benthic 'protozoans' (more correctly the animal-like protists) that pass through a 0.5-mm mesh and, by extension, the same term is often used for heterotrophic protists from other benthic habitats, although it is not used in respect of the PLANKTON. [R.S.K.B.]

microflora (symbionts) Usually refers to microscopic algae and possibly also bacteria and fungi. *Cf.* MICROFAUNA. [P.C.]

microgeographic variation Phenotypic or genetic variation among patches in a local HABITAT that may arise from migration, behavioural selection for appropriate substrates or selective MORTALITY among GENOTYPES. [V.F.]

microhabitat Precise physical location of a species. [P.C.]

micronutrients These are elements required by living organisms only in small amounts (*cf.* MACRONUTRIENTS). The great majority, for example boron, manganese, copper, zinc and molybdenum in plants and fluorine, manganese, cobalt, copper, zinc, selenium and iodine in animals, are trace elements; iron is also an essential micronutrient even though it is very abundant in the LITHOSPHERE. [K.A.S. & J.R.E.]

microtine cycles The enormous changes in population abundance characteristic of certain small rodents, particularly voles and lemmings, that lead to regular population explosions and crashes. The cycles occur with a fairly regular periodicity of between 2 and 5 years, but may vary widely in amplitude. [V.F.]

midoceanic ridge The longest mountain chain on Earth, extending through the Atlantic, Indian, Antarctic and South Pacific oceans, the Norwegian Sea and the Arctic Basin to a distance of more than 56 000 km. The midocean ridges are areas in which new crustal material is forming as regions of the Earth's outer shell move apart. [V.F.]

mimicry Mimicry refers to striking similarities in the colour patterns and morphology of different species. Two fundamental forms of mimetic resemblance have long been recognized. Müllerian mimicry occurs between two or more species (called a mimicry ring), each of which has a closely similar warning, or aposematic, colour pattern. Their colour patterns act as conspicuous signals to visually hunting predators of some form of toxicity, noxiousness or distastefulness.

The second major form of mimicry is BATESIAN MIMICRY. In contrast to the mutual benefit derived by the protected, unpalatable co-mimics within a Müllerian relationship, Batesian mimicry involves one (or more) palatable species exhibiting the same colour pattern as an unpalatable species. The latter species acts as a model for the former, which is the mimic. Although the mimic is perfectly palatable to birds or lizards, it derives some protection because potential predators sometimes mistake it for the unprofitable prey they have previously learnt to avoid. The phenomenon is of interest to evolutionary biologists partly because there is a conflict. This arises because while the mimic derives some protection in the presence of the model, the model species is more likely to be sampled by predators when a Batesian mimic is present. The degree of protection derived by the mimic is also frequency dependent: the more abundant the mimic relative to the model, the less well protected the mimic. [P.M.B.]

mineral nutrients Those NUTRIENTS that originate mainly by weathering of minerals in the SOIL, including the MACRONUTRIENTS nitrogen, phosphorus, potassium, calcium, magnesium and sulphur, and the MICRONUTRIENTS iron, manganese, boron, molybdenum, zinc, chlorine and cobalt. [J.G.]

minimum viable population (MVP) This is a popular concept in CONSERVATION BIOLOGY and is defined as the threshold POPULATION SIZE below which rapid extinction is virtually guaranteed. Clearly, in general, smaller populations will be more vulnerable than larger ones, although it has not been established, either theoretically or empirically, that there is some critical population size below which the vulnerability to extinction increases suddenly. It is perhaps more useful to estimate extinction probability as a function of time for different population sizes than to identify some specific MVP.

MVP size has been defined as that which would have at least a 99% chance of surviving 1000 years. It has also been suggested that 'endangered' could be defined as a 20% chance of extinction in 20 years. *See also* ENDANGERED SPECIES; EXTINCTION MODELLING. [P.C.]

mire This term has been used in a variety of ways, some specific and many broad and general. Over the past 30 years it has been used increasingly of PEAT-forming habitats and ecosystems and has proved a most valuable general term in this area. [P.D.M.]

mobbing The close attendance and alarm behaviour by certain vertebrates directed towards potential predators that serves to distract the attention of the PREDATOR and direct the attention of conspecifics and heterospecifics to the predator. [A.P.M.]

mode This MEASURE OF CENTRAL TENDENCY or average is simply the value of the most common observation. It is the 'peak' of the DISTRIBUTION. [B.S.]

model A simplified representation of the real world that aids understanding. In ecology, models of the real world are usually constructed from equations (*see* LOTKA–VOLTERRA MODEL). [B.S.]

models of parasitoid–host interaction The first models describing the population-dynamic interactions between PARASITOIDS and their HOSTS were developed by Thompson and Nicholson in the 1920s and 1930s. While Thompson thought that parasitoids were primarily limited by their egg supply, Nicholson considered host limitation to be more important and it is his ideas that have proven more influential. The Nicholson–Bailey model applies to hosts and parasitoids with discrete, perfectly synchronized generations, as is often found in temperate parasitoid–host interactions. The model assumes that parasitoids search randomly and that the average number of parasitoid encounters per host over the season is the product of parasitoid density (P) and a scaling constant (a) normally called the SEARCHING EFFICIENCY or attack rate. The proportion of hosts escaping PARASITISM is given by the zero term of a POISSON DISTRIBUTION with mean aP (which is $\exp(-aP)$). So the proportion of hosts parasitized is simply $1 - \exp(-aP)$. The numbers of hosts in the next generation is given by the number of hosts escaping parasitism multiplied by net host FECUNDITY. The number of parasitoids in the next generation is given by the number of parasitized hosts multiplied by a constant representing the number of parasitoids emerging per host. The Nicholson–Bailey model is a coupled difference equation whose equilibrium and stability properties are easy to analyse. The model always has an UNSTABLE EQUILIBRIUM and predicts diverging OSCILLATIONS. That the simplest host–parasitoid model does not predict a persistent interaction has led to much research on factors that may stabilize this type of interaction.

Functional response and interference. Early work focused on the parasitoid FUNCTIONAL RESPONSE and INTERFERENCE. Type 1 and type 2 functional responses tended to increase the tendency for diverging oscillations, as they weaken the ability of the parasitoid to control an expanding host population. Type 3 functional responses could contribute to stability, as over some ranges of host densities parasitoids respond to increasing host densities by an accelerating rate of host parasitism. However, for most realistic type 3 functional responses, their effect on stability is likely to be small. Interference is a reduction in parasitoid PER CAPITA efficiency with increasing parasitoid density. It exerts a stabilizing influence because it acts to prevent high densities of parasitoids from overexploiting their host population. Although interference can be demonstrated in laboratory experiments, most workers today believe that interference is unlikely to be strong enough in the field to be a major influence on host–parasitoid POPULATION DYNAMICS.

Heterogeneity of risk. The major factors contributing to persistent host–parasitoid interactions are currently considered to revolve around processes that tend to contribute to increased variance in the risk of parasitism across host populations. If some hosts, for whatever reason, are protected or partially protected from parasitoid attack, then the host population will suffer less catastrophic MORTALITY when parasitoid densities are high and this acts to damp the tendency for the system to display diverging oscillations. A number of factors have been shown to promote stability by increasing the variance in host risk.

1 Some hosts may exist in a physical refuge from parasitoid attack.

2 Some hosts may have the ability to defend themselves physically or physiologically from parasitism.

3 Hosts may be distributed heterogeneously across patches in the environment and parasitoids may tend not to visit certain patches.

4 The host and parasitoid populations may not be perfectly synchronized in time so that some hosts enjoy a reduced risk of parasitism.

Spatial processes. Recent work has begun to consider spatial models of host–parasitoid interactions. A metapopulation of individually unstable subpopulations (for example described by the Nicholson–Bailey model) connected by migration can be persistent. As long as different subpopulations oscillate out of synchrony then a rescue effect is possible, whereby a particular subpopulation near extinction can be rescued by migration from out-of-phase subpopulations. Any factor that tends to synchronize subpopulations (external environmental forcing, very frequent migration) leads to instability, as does very low levels of migration which destroys the rescue effect.

Age-structure. A second recent trend has been the development of continuous-time models of host–parasitoid interactions, which assume overlapping as opposed to DISCRETE GENERATIONS. The continuous time equivalent of the Nicholson–Bailey model is the LOTKA–VOLTERRA MODEL, which predicts neutral as opposed to divergent oscillations. However, the Lotka–Volterra model implicitly assumes no host and parasitoid developmental lags: their incorporation into a modified Lotka–Volterra model (expressed as time-delayed differential or integral equations) leads to dynamic behaviour that has much in common with their discrete generation analogues. One new feature predicted by these models is the stabilizing influence of long-lived host stages that are immune from parasitoid attack. These 'storage stages' act as a refuge, allowing the host population to survive periods of heavy parasitoid attack. Where hosts and parasitoids have different developmental periods, the two time-lags can interact dynamically to give persistent cycles. [H.C.J.G.]

models of predator–prey interaction Symbolic abstractions of natural, exploitative relationships between species. The exploiter is the PREDATOR or enemy; its victim is the prey. Predator–prey models allow deductions about the POPULATION DYNAMICS and POPULATION SIZES of the interacting species.

Alfred Lotka and Vito Volterra independently produced the first mathematical model of predator–prey interaction in the 1920s. They stripped away all but four of the two species' natural features: *b*, PER CAPITA birth rate of prey; *d*, per capita death rate of predators; *k*, predation rate; and β, per capita birth rate of predators per predation event. Then they combined the birth and death rates of each species by assuming their additivity and wrote a pair of simultaneous DIFFERENTIAL EQUATIONS:

$$\mathrm{d}V/\mathrm{d}t = dV - kVP$$

$$\mathrm{d}P/\mathrm{d}t = \beta kVP - dP$$

where *V* and *P* are the population densities of predator and victim, respectively. In this model, notice that in the absence of the predator, the prey density grows infinitely, whereas in the absence of the prey, the predator becomes extinct. Inadequate though it may be, Volterra used this model to produce a robust theory of BIOLOGICAL CONTROL.

A much better model results from changing the predator FUNCTIONAL RESPONSE (i.e. the rate at which an individual predator kills prey). A non-linear response models predatory HUNGER and SATIATION. There are two alternatives.

1 Half-saturation model: $kV/(\chi + V)$, where χ, the half-saturation constant, is the victim density at which a predator kills at $k/2$, i.e. half its maximum rate. This response model was conceived independently at least three times. Its most popular form, HOLLING'S DISC EQUATION, is known as the type 2 response, although the type 2 and the half-saturation forms are mathematically identical. They are also identical to the Michaelis–Menten equation of enzyme chemistry.

2 Ivlev model: $k(1 - e^{-cV})$, where *c* is a constant proportional to the predator's search speed.

These models introduce the possibility of an UNSTABLE EQUILIBRIUM.

Many other algebraic elaborations of predation models attempt to reconcile empirical work with theory. Some of these concentrate on particular interactions, like sheep and blowflies. Some are more general.

Andrei Kolmogorov introduced time-lags to predator–prey models. These make reproduction sensitive to preceding rather than current conditions. Time-lags can introduce instability to predator–prey dynamics.

Predation models may result in an additional dynamical behaviour called CHAOS. Chaotic equilibria, belying their name, are fully deterministic, but trajectories around them are so dependent on initial conditions and so variable from oscillation to oscillation that they resemble those dominated by stochastic influences. [M.L.R.]

modular organism A modular organism has an indeterminate morphological form that changes throughout its LIFESPAN. It is constructed from a number of basic subunits (e.g. reproductive and vegetative plant GROWTH), which in turn produce further 'clonal' copies. The exact form (e.g. number of branches) is strongly influenced by the surrounding environment in which the organism develops. Physiologically, independent subunits are known as ramets, while the whole 'genetic individual' is known as a GENET. Such organisms are typically SESSILE in their adult stages. Although plants are the most obvious examples of modular organisms, there are also a number of modular animals such as corals and bryozoans. [R.C.]

module Genetically identical parts of a modular (clonal) individual (the GENET) that might be used as a basis for counting when assessing the ABUNDANCE of an organism (e.g. the number of tillers of a grass plant or the number of workers in an ant colony). [M.J.C.]

molecular ecology Molecular ecology is a generic term loosely applied to ecological studies that employ the tools of molecular biology. [P.B.R.]

monitoring Continuous or regular assessment of the quality of emissions or EFFLUENTS, or sources that lead to these, or specific places in the environment possibly subjected to them. For the latter, this can involve both chemical and biological assessment. *See also* POLLUTION. [P.C.]

monoculture The agricultural, arboricultural and horticultural practice of cultivating one plant species, often a single cultivar, in one location. [G.A.F.H.]

monoecy The reproductive process where separate male and female flowers are developed on the same plant (*see also* HERMAPHRODITE). [G.A.F.H.]

monogamy A mating system in which a single male and female mate only with each other during a breeding season, common in birds but rarer in mammals and other animals. [J.C.D.]

monophagous Monophagous organisms are only able to exploit a single type of food resource. This may consist of either a single species or a number of closely related species that provide very similar food types. Monophagy is common between species that have been subjected to prolonged periods of COEVOLUTION or where habitat fragmentation has produced associated 'host shifts'. [R.C.]

monotonic damping The complete lack of fluctuations in POPULATION SIZE as the population approaches its equilibrium. The lack of fluctuation results from perfect COMPENSATION of density-dependent responses, for example as population density increases, individual reproductive output or survival decreases at a rate that exactly compensates for the increased density. [V.F.]

monotypic Taxa (often families) having only a single species. [P.C.]

monsoon forest Tropical and SUBTROPICAL forests of certain areas, such as northern India and Burma, are supplied with rainfall during a wet season as a result of monsoon rains penetrating the continent as warm, moist winds blow from the ocean. During the wet season growth is lush, but in the following DRY SEASON the forest may suffer drought. During the drought many trees lose their leaves, thus avoiding excessive transpiration. [P.D.M.]

montane forests High-elevation forests are those forests below the TIMBER LINE (subalpine) but at least 1000 m (3280 feet) above sea level. These elevations apply generally, although higher elevations are required at lower latitudes before montane forests can occur. [P.M.S.A. & G.P.B.]

Monte Carlo simulation Method used to explore UNCERTAINTY in mathematical simulation models. It involves an iterative process in which model parameter values are selected at random from a specified frequency distribution. Output values, following simulation runs, are used to determine the PROBABILITY of occurrence of particular values, given the uncertainty in the parameter. [P.C.]

moorland Upland areas of heather-dominated vegetation in north-western Europe are termed 'moorland'. The floristic distinction between HEATHLAND and moorland is imprecise. [P.D.M.]

mor A surface HUMUS horizon that forms beneath conifers and open HEATHLAND and MOORLAND in temperate climates. It is acid, low in most microbial activities (not of fungi) and consists of several separated layers in various degrees of decomposition. [P.C.]

morbidity Any departure from a state of physiological (and/or psychological in people) well-being. Synonymous with sickness, illness or ill health. It is more difficult to measure than MORTALITY but can be done, for example in terms of proportion of population 'ill', frequency of periods of illness, duration of illness. [P.C.]

mortality At its most simplistic level, mortality describes the number of individuals dying in a particular population. [R.C.]

mudflat 'Mud' is not a precise concept: it generally denotes any ground that is soft and wet. Likewise 'mudflat' is applied to any TIDAL FLAT that does not have the usual clean appearance of sand. In practice, this means that appreciable quantities of SILT (particles 0.002–0.06 mm diameter) are present even though the major sedimentary component may still be sand (0.06–2.0 mm diameter) or even gravel (2–60 mm diameter). 'Sandy mud' and 'muddy sand' are terms often used to indicate muddy sediments dominated by silts or sands, respectively. [R.S.K.B.]

mull A surface HUMUS that forms in deciduous forests. It is neutral or alkaline. It provides generally favourable conditions for decomposition and in consequence it is well mixed. [P.C.]

Müllerian mimicry *See* MIMICRY.

multiparasitism The deposition of one or a clutch of EGGS by a PARASITOID on a HOST that has already been parasitized by a member of a different parasitoid species. [H.C.J.G.]

multiple equilibrium Alternative stable states in the ABUNDANCE of predators and their prey. [V.F.]

multiple stable states At any given time, a dynamic system is described by a point in its state space. In the course of time, this point moves on a trajectory through state space as the state of the system changes due to the forces that act upon it. A stable state is represented by a point in state space that is stationary. In such a state, the sum of the forces acting on the system is zero and the state of the system remains constant. A stable state is also called an equilibrium. The system has multiple stable states if there is more than one stationary point in its state space. [M.D.]

muskeg The PEATLANDS of Canada are often referred to as muskeg. They have been classified into the confined peatlands of the southerly areas and the unconfined cover of peatland in the higher latitudes. [P.D.M.]

mutation In his book *The Mutation Theory*, De Vries distinguished between continuous, individual variation and discrete saltational variations. He applied the term to the latter only. Now the term is used for any spontaneous, random change, large or small, in the genome. However, a distinction is sometimes made between a point mutation (impact on a single gene or even base substitution) and a macromutation (change with major, usually obvious visible, effect on the PHENOTYPE). Random mutation is intended to imply non-directional with respect to the course of EVOLUTION. Mutation generates variety upon which NATURAL SELECTION operates. [P.C.]

mutualism A widespread form of interaction between individuals of two or more species that results in mutual benefit (sometimes abbreviated to a ++ interaction). Where the participants are in close physical contact the mutualism can be described as a symbiosis. Symbiotic mutualisms are typically OBLIGATE, with one partner unable to persist in the absence of another, whereas non-symbiotic mutualisms are often FACULTATIVE, with only a proportion of the individuals of a species involved. [S.G.C.]

mycelium The fungi are characterized by a LIFE CYCLE in which their dispersed propagules (spores) germinate on a favourable substratum by formation of a germ tube, which develops into a radiating system of filamentous, branching, tubular threads termed 'hyphae', producing a three-dimensional network collectively known as the mycelium (representing the thallus). [J.B.H.]

myrmecochory The dispersal of seeds by ants. [M.F.]

myrmecophile A plant (or animal) that lives in close association with ants. [M.F.]

N

natality (fecundity) A measure of the production of offspring in a population. As with MORTALITY, this major demographic factor may also be expressed in a number of different ways. Three of the most commonly used measures are the BASIC REPRODUCTIVE RATE (R_0), the net reproductive rate (R) and the INTRINSIC RATE OF INCREASE (r). R_0 describes the average number of offspring produced within a particular age-specific group; R describes the fundamental PER CAPITA rate of increase in a population (i.e. numbers born relative to the numbers surviving); and r describes the continuous rate of change in a population. Both the basic reproductive rate and the net reproductive rate are discrete measurements that are usually made over long time intervals, for example annually. In contrast, the intrinsic rate of increase is an instantaneous measurement which is usually made over shorter periods. There is a general relationship between these different quantities where:

Instantaneous rate = Natural log of finite rate

i.e.:

$r = \ln R$

[R.C.]

native and naturalized species Native species are those that have evolved in place or have become established independently of human activity. Naturalized species were brought to a place by human activity, either intentionally or not, and have subsequently established self-sustaining populations. [E.O.G. & L.R.M.]

natural Literally, of nature. The term usually implies a situation in which there has been no modification by humans: for example, a natural landscape, a natural community. [P.C.]

natural classification Any CLASSIFICATION that contains only monophyletic groups and is thus logically consistent with the evolutionary relationships of the organisms. [D.R.B.]

natural disasters Upheavals leading to impacts on human society and/or ecological systems and/or the abiotic environment by phenomena that are not attributable to human causes. Examples include earthquakes, volcanoes, storms, tornadoes, hurricanes, floods, forest fires, DROUGHT and meteorite impacts. [P.C.]

natural selection A process involving differences between individuals in their rates of survival or reproduction, leading to some types of organism being represented more than others in the next generation. The process of natural selection is generally held to explain most if not all evolutionary change. [J.R.G.T.]

naturalized Describing species that have established themselves and are flourishing following their introduction to a non-native region. [P.O.]

nature versus nurture Shorthand for the relative involvement of, respectively, heredity (genetics) and ENVIRONMENT in the control and development of TRAITS. [P.C.]

nearest neighbour techniques There are several nearest neighbour techniques available and they may be used both to estimate POPULATION SIZE and to describe the type of DISTRIBUTION of a population. These techniques usually require selection of random individuals and then determination of the distance from that individual to its nearest neighbour. [C.D.]

necromass Dead organic matter accumulated in a COMMUNITY, *cf.* BIOMASS. [P.O.]

necrotrophic Describing PARASITES that are able to continue to live on a dead HOST. [P.O.]

negative binomial distribution A discrete PROBABILITY DISTRIBUTION that describes situations where observations (e.g. individuals in a SAMPLING unit) follow an AGGREGATED DISTRIBUTION in either space or time. [B.S.]

negative feedback The coupling of an output process to an input process in such a way that the output process is inhibited. [S.N.W.]

nekton Pelagic animals whose swimming abilities permit them to move actively through the water column and to move against currents. They include small crustaceans, cephalopods, fish, reptiles, birds and marine mammals. Nektonic species show adaptations for maintaining their buoyancy (e.g. gas or swim bladders in fish) and for effectively moving through water (e.g. a streamlined body form, paddle-shaped appendages, undulatory body movements). [V.F.]

neotropical The biogeographical region which comprises the tropics and subtropics of the New World (in contradistinction to palaeotropical, i.e. of the Old World tropics). It has been taken to include not only the Caribbean region, southern Mexico, southern Florida, Central America and the northern part of

South America, but also extending as far south as Argentina and Chile, far outside the tropics. [W.G.C.]

neritic Referring to inshore waters between mean low water to 200 m in depth that overlie CONTINENTAL SHELVES (i.e. the zone landward of the shelf-slope break). The neritic zone is characterized not by a fixed depth but by the fact that it is euphotic and thus an area in which PHOTOSYNTHESIS can occur. [V.F.]

nest parasitism The exploitation of dependent offspring in nests by invertebrate parasites that use blood or other tissue of the offspring as a source of nutrition (*cf.* BROOD PARASITISM). The nests of insects, birds and mammals often contain large numbers of invertebrates that either exploit the nest, its contents or its owners as a resource. Nest parasites include a large number of species of mites and ticks, fleas, lice, louseflies and blowflies. [A.P.M.]

net primary production (NPP) The process of energy, matter or carbon accumulation net of respiratory losses by the AUTOTROPHS of a community. Its rate, net primary productivity (P_n or N), equals the difference between gross primary productivity and RESPIRATION by the autotrophs. Although the term applies to all autotrophs it is most commonly used in the context of photosynthetic organisms. [S.P.L.]

net rate of increase The multiplication rate per generation of a population (also known as the 'net reproductive rate'), defined as the ratio of total female births in successive generations and usually designated R, or R_0. R is calculated from a COHORT LIFE TABLE containing age-specific MORTALITY and FECUNDITY schedules. If l_x is the proportion of females surviving to age x and m_x is their fecundity (female offspring) at that age, then the total number of female offspring produced during a single generation is the sum of the $l_x m_x$ products over the reproductive life of females:

$$R = \Sigma l_x m_x$$

[R.H.S.]

net recruitment curve A graph of RECRUITMENT in relation to another factor, in particular in relation to the POPULATION DENSITY. In general, the curves are 'n'-shaped since when density is low NATALITY is usually high and MORTALITY is low, but when density is high natality is low and mortality is high; however, other forms are possible. Real, or theoretical, net recruitment curves are used to estimate the MAXIMUM SUSTAINABLE YIELD possible when HARVESTING organisms from a population. [A.J.D.]

neuston Planktonic organisms of the PELAGIC zone that live at or very near the surface. [V.F.]

neutral theory of evolution The neutral theory of EVOLUTION considers that the majority of the genetic differences between individuals or between species are due to the accidental accumulation of mutant DNA sequences that have little or no effect on individual reproductive success. [G.A.D.]

neutralism A situation in which two species, living in the same habitat, have no effect upon each other (0 0 interaction). *See also* SPECIES INTERACTIONS. [B.S.]

niche This term is often used loosely by both ecologists and laymen, but defining the niche concept formally is rather complicated. An old interpretation considers it to be the functional role or 'occupation' of a species in a COMMUNITY. Presently, the generally used concept is that of the 'Hutchinsonian niche' (after G.E. Hutchinson: *see* CHARACTERS IN ECOLOGY), which is defined as the requirements that the ENVIRONMENT has to meet to allow the persistence of (a POPULATION of) a species. The requirements consist of the tolerable ranges of conditions, i.e. those characteristics of the environment that are not influenced by the organism (such as salinity), and the tolerable ranges of RESOURCES that are depletable (such as food types and nesting sites). The distinction between conditions and resources is not always clear. Light and carbon dioxide (CO_2) may be conditions for a solitary plant, but may be resources in dense vegetation.

The tolerable ranges of conditions and resources are each envisaged as a section of an axis of a multidimensional 'niche space'. Together, the sections define the niche as a multidimensional hypervolume in niche space. Inside the hypervolume, the requirements for persistence of the species or population are met. The requirements along different axes may interact. For example, a high temperature may decrease the upper tolerance limit for salinity, or one type of food may be replaced by another. Such cases will produce rounded or flattened corners of the hypervolume. [J.G.S.]

niche overlap Quantification of the similarity of niches and, in particular, of the similarity in the use of resource types. The RESOURCES involved can be food, but also, for instance, may be aspects of the HABITAT. Many measures of NICHE overlap have been proposed, some of which are similar to the measures used to express the similarity of COMMUNITIES. [J.G.S.]

niche width (niche breadth) Quantification of the degree of specialization of a species (or a POPULATION), i.e of the degree to which different resource types are used. Many measures of NICHE width have been proposed, some of which are similar to measurements of DIVERSITY in communities. [J.G.S.]

Nicholson–Bailey equations A discrete-time development of the Lotka–Volterra predator–prey model (*see* LOTKA–VOLTERRA MODEL), usually couched in terms of the numbers of hosts and parasitoids. The model was developed by the Australian entomologist A.J. Nicholson and physicist V.A. Bailey in 1935.

Let H_t be the number of hosts at time t, and P_t the number of parasitoids. The Nicholson–Bailey equations are:

$$H_{t+1} = H_t\lambda \exp(-aP_t)$$

$$P_{t+1} = H_t[1 - \exp(-aP_t)]$$

where λ is the net per capita FECUNDITY of the hosts and a is a measure of the parasitoids' SEARCHING EFFICIENCY. The model assumes parasitoids and hosts have discrete OVERLAPPING GENERATIONS and that parasitoids search randomly for their hosts. The model predicts diverging OSCILLATIONS in population densities and is an important normative case with which more complicated and more realistic models are compared. [H.C.J.G. & R.M.S.]

nidicolous (nest-squatters) The term nidicolous (from the Latin *nidus* nest, and *colere* to inhabit) is used to describe the nest attendance of neonates of birds and mammals. The contrasting terms nidicolous and NIDIFUGOUS (from the Latin *fugere* to flee) were introduced by Oken in the early 1800s and referred to *Nestflüchter* (nest-fleers, or nidifugous birds) and *Nesthocker* (nest-squatters, or nidicolous birds). Originally, nidicolous and ALTRICIAL were used as synonyms, as were nidifugous and PRECOCIAL. Today, these pairs of terms are used in slightly differing contexts although their general meanings overlap in some aspects. Altricial and precocial primarily refer to the developmental stage of the chick, whereas nidicolous and nidifugous refer to nest attendance. [J.M.S.]

nidifugous (nest-fleers) A term referring to the developmental state of neonates in mammals and birds. *Cf.* PRECOCIAL. *See also* NIDICOLOUS. [J.M.S.]

nitrogen cycle The movement of nitrogen between the ATMOSPHERE, the BIOSPHERE and the HYDROSPHERE, and the associated transformations between different chemical forms. [K.A.S.]

non-equilibrium communities In essence the NON-EQUILIBRIUM THEORY of COMMUNITY organization is predicated on the idea that community structure is primarily determined by the vagaries of the environment such that density-independent controls operate and that COMPETITION between species is minimal. In contrast, the importance of competition is central to the equilibrium viewpoint where communities are considered to be saturated with species, which partition all of the available RESOURCES (*see* EQUILIBRIUM, COMMUNITIES; SATURATION). [S.J.H.]

non-equilibrium theory The theory of an interaction where stable end-points, or equilibria, are not predicted but where the system changes constantly. [P.I.W.]

non-indigenous species A species occurring outside its NATURAL geographical RANGE; also known as an ALIEN species, in contrast to a native species (*see* NATIVE AND NATURALIZED SPECIES). [M.J.C.]

non-interactive community A COMMUNITY in which interactions between species at the same TROPHIC LEVEL within a local habitat (*see* SCALE) are weak or absent. Although most emphasis in ecology is on interactive communities there is evidence that, for some groups, non-interactive communities may be more common. [S.J.H.]

non-linear models Models in which the mean response is not a linear combination of the input PARAMETERS. For example, the model $Y = \alpha e^{\beta x} + \varepsilon$ is non-linear in its parameters since the mean response, Y, is an exponential, rather than a linear, function of the parameter β. [V.F.]

non-Mendelian inheritance This most commonly refers to the transmission of continuously varying (also called quantitative, metric or polygenic) TRAITS from one generation to the next. This is in contrast to the transmission of traits that fall clearly into distinct phenotypic classes (i.e. Mendelian traits). Whereas Mendelian traits are controlled by single genes having large phenotypic effects, non-Mendelian traits are controlled by many genes each having a small phenotypic effect. [V.F.]

non-parametric statistics Statistical tests are frequently classified into parametric and non-parametric tests. Non-parametric ones do not require quantitative information/observations to follow a NORMAL DISTRIBUTION and to have similar VARIABILITY (homogeneous variances). Non-parametric tests can be carried out on qualitative information, i.e. ranked information or information in categories. [B.S.]

non-target species Any species that is adversely affected by the application of a BIOCIDE that is not intended to affect that species. [P.C.]

normal distribution A continuous PROBABILITY DISTRIBUTION that is symmetrical, unimodal and bell-shaped. The equation defining the shape and position of the curve is:

$$y = \frac{1}{\sigma\sqrt{2\pi}} \exp\left[-\frac{1}{2}\left(\frac{x-\mu}{\sigma}\right)^2\right]$$

where x is the measured variable, μ is the MEAN and σ is the STANDARD DEVIATION. The notation exp[ ] represents $e^{[\,]}$, with e the base of natural, or Napierian, logarithms. Both e (= 2.71828) and π (= 3.14159) are constants. The shape and position of any normal distribution is therefore defined by the mean and standard deviation (Fig. N1). For all normal distributions, the following percentages of observations lie within the indicated number of standard deviations, either side of the mean:

68.26% of observations within the $\mu \pm \sigma$

95.44% of observations within the $\mu \pm 2\sigma$

99.73% of observations within the $\mu \pm 3\sigma$

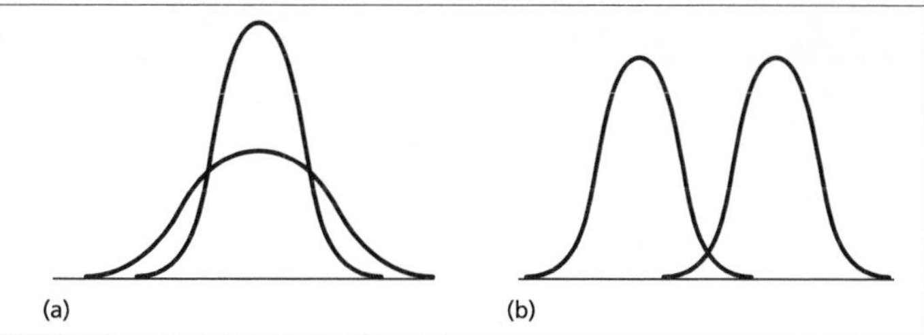

Fig. N1 Normal distributions. (a) Same μ different σ. (b) Different μ same σ.

More conveniently:

95% of all observations lie within the $\mu \pm 1.960\sigma$

99% of all observations lie within the $\mu \pm 2.576\sigma$

Therefore, in a normal distribution, 5% of all observations fall outside the limits of $\mu \pm 1.960\sigma$, and 1% of all observations fall outside the limits of $\mu \pm 2.576\sigma$. These limits are important when carrying out statistical tests, such as the *F* TEST and *t* TEST. *See also* DISTRIBUTION. [B.S.]

nuclear winter A term coined to express the effect of the increased smoke and dust burden in the ATMOSPHERE which would result from the explosion of thermonuclear devices at the surface and in the air as a result of a nuclear war. [J.B.M.]

nuisance species Non-indigenous species, usually introduced as a result of human activity, that threaten the density and/or abundance of native species. [P.C.]

null hypothesis (*H_0*) The hypothesis that forms the basis of a statistical test. A neutral, or null, hypothesis is one proposing no difference, and is usually symbolized by H_0. For example, when carrying out a *t* TEST, the null hypothesis is that the two SAMPLE means ($\bar{x}_1$ and $\bar{x}_2$) are from populations with the same means (H_0: $\bar{\mu}_1 = \bar{\mu}_2$); i.e. if the sample means are different it is only due to chance. Similarly, when carrying out an *F* TEST, the null hypothesis is that the two sample variances ($s_1{}^2$ and $s_2{}^2$) are from populations with the same variances (H_0: $\sigma_1{}^2 = \sigma_2{}^2$); i.e. if the sample variances are different it is only due to chance. There is always an alternative hypothesis to H_0. [B.S.]

null model A term used in the investigation of COMMUNITY structure, and used mainly in COMMUNITY ECOLOGY and comparative analyses for a more precise, usually quantitative, NULL HYPOTHESIS. [P.I.W.]

numerical response The relationship between the number of predators or parasitoids in an area and the DENSITY of their prey or HOST. The numerical response may consist of a short-term behavioural and a longer-term reproductive component. The numerical response and the FUNCTIONAL RESPONSE act together to determine rates of predation. [H.C.J.G.]

numerical taxonomy Although this term might be applied to any numeric approach to TAXONOMY and CLASSIFICATION, it is now often taken to refer primarily or even exclusively to those numerical methods that group individuals or taxa on the basis of some function of their similarity to one another—often referred to as distance methods or phenetics. This contrasts with the cladistic approach. [D.L.J.Q.]

nutrient budget The nutrient budget, or the nutrient balance, is an attempt to account quantitatively for the inputs ('income') of NUTRIENTS to an ECOSYSTEM, and the losses. In natural ecosystems, income is from rain, wet and dry deposition, nitrogen fixation and weathering of rocks. [J.G.]

nutrients Raw materials needed for life. Their exact nature varies according to whether the living system is heterotrophic or autotrophic—and even varies between species within these groupings. The term can include water, minerals, inorganic salts and various organic molecules (proteins, carbohydrates, fats and vitamins). [P.C.]

O

obligate Describing a factor that must be present in order for an organism to survive in a particular environment. When the requirement is obligate then survival is not possible if the factor is not present; for example, some bacteria can only be cultured in conditions that exclude oxygen, and are consequently described as obligate anaerobes. The definition most commonly refers to the physical rather than the biological features of an environment. [R.C.]

ocean The vast body of water covering 70.8% of the surface of the globe, or any one of its principal divisions: the Antarctic, Atlantic, Arctic, Indian and Pacific Oceans. The average depth of the ocean is 3729 m and the deepest parts are 11 022 m from the ocean surface. The total area of the world ocean is 3.62×10^8 km^2, and its volume is 1.35×10^9 km^3. The ocean is believed to have formed between 4400 and 3000 million years ago. All known phyla originated in the ocean. [V.F.]

old-growth/ancient forests The term 'old-growth' describes forests that have developed over a long period of time without experiencing catastrophic DISTURBANCE. The age at which old-growth develops and the specific structural attributes that characterize old-growth vary with FOREST type, climate, site conditions and disturbance regime. For example, an alder forest may attain old-growth characteristics at 80 years, while a fir forest might not be considered old-growth until age 200 years or more. While old-growth has been defined simply on the basis of tree age, size or growth rate, most ecologists use a combination of structural characteristics for distinguishing old-growth from younger forests (not all of which may be present in a given forest):

- large old trees (for a species and site);
- trees with broken or deformed tops, or with stem and root decay;
- wide variation in tree size and spacing;
- tree foliage well distributed vertically;
- large dead trees (both standing and on the ground) in a range of decay states;
- CANOPY openings from tree-fall gaps;
- patchiness of understorey vegetation.

[A.N.G.]

oldfield ecosystems Extensive areas of abandoned farmland in the north-east USA that have been recolonized in the past half-century by now-maturing trees. [G.A.F.H.]

oligophagous Describing herbivorous organisms, typically insects, that utilize a range of food RESOURCES. [R.C.]

ombrogenous Describing peatland that has developed under conditions where the sole input of water (and therefore nutrient elements) is derived from precipitation falling directly upon it. Literally, originating under the influence of rainfall. [P.D.M.]

ombrophilous Describing habitats, vegetation or individual species that require a water source (and therefore nutrient element source) that is solely derived from direct precipitation; literally, 'rain-loving'. [P.D.M.]

ombrotrophic Describing peatland that derives its nutrient input entirely from the atmosphere via direct precipitation; literally, 'rain-fed'. *See also* MIRE. [P.D.M.]

omnivore An animal that feeds on both animal and vegetable matter. [P.O.]

one-tailed test In the NORMAL DISTRIBUTION 95% of observations are found to one side of $\mu \pm 1.65\sigma$ (where μ is the MEAN and σ is the STANDARD DEVIATION) and the other 5% of observations are found in the remaining tail of the distribution. When we carry out a $z(d)$ TEST to examine the difference between two means, and we have a priori reasons to expect a value of the test statistic above or below $z = 0$ (one mean should be larger), we use the 5% level of significance that corresponds to $z = 1.65$. This is a one-tailed test, and corresponds to a NULL HYPOTHESIS (H_0): $\bar{x}_1 = \bar{x}_2$ and an alternative hypothesis (H_1): $\bar{x}_1 > \bar{x}_2$. Any statistical test that uses the PROBABILITY in one tail of a distribution is therefore a one-tailed test. The alternative is to use the value of the test statistic that cuts off a total of 5% in both tails. For the normal distribution this would correspond to $z = 1.96$. Using this value as the 5% significance point would imply a TWO-TAILED TEST. [B.S.]

ontogeny The LIFE CYCLE of an individual, both embryonic and postnatal. [P.A.]

open canopy Stands of trees in woodlands and forests in which the density of trees in the upper stratum is low enough to allow significant gaps to be present between the crowns of neighbouring trees. [G.F.P.]

open community An ASSEMBLAGE of organisms with incomplete ground COVER (*cf.* CLOSED COMMUNITY), readily invaded by immigrant organisms, some ecological niches being unoccupied. In open VEGETATION, gaps exist in the CANOPY which may separate the plants by up to as much as twice the diameter of the predominant individuals. If the vegetation is even more scanty—the ground space exceeding the plant cover—the vegetation may be termed 'sparse'. Open communities are characteristic of early stages of SUCCESSION. [A.J.W.]

open system An open system is commonly used to describe a physical system in which energy and matter can enter and leave without restriction. The definition can, however, also be applied to biological systems. In a physical sense the BIOSPHERE is an open system when radiant energy is considered but closed when geochemical cycles (e.g. carbon) are considered. [R.C.]

opportunistic species Species that can successfully exploit new RESOURCES as and when they arise. Such species are also termed COLONIZING SPECIES, RUDERAL species or *r*-selected species (*see* *r*- AND *K*-SELECTION). Conditions suitable for the growth of opportunistic species occur unpredictably in time and space and last only for several generations. [T.J.K.]

optimal age at maturity In evolutionary biology, age at maturity is typically defined as the age at first reproduction. Coupled with SIZE AT MATURITY, it is an important FITNESS component, usually under STABILIZING SELECTION with an optimum determined by the balance of costs and benefits of delayed maturity. [T.J.K.]

optimal clutch size The number of offspring produced per breeding attempt that maximizes FITNESS. The optimal CLUTCH SIZE is a function of TRADE-OFFS between FECUNDITY and survival in the parents, and SURVIVAL RATE of offspring resulting from changes in provisioning due to variation in clutch size and/or earlier breeding patterns of the parents. [D.A.R.]

optimal control theory Optimal control theory (or dynamic optimization theory) is a body of theory and computational techniques that is useful for solving dynamic optimization models. It is an essential ingredient in much of modern technology, particularly in the aerospace and communications industries. In evolutionary biology optimal control theory is being used increasingly in the study of behavioural and life-history ADAPTATIONS. A particular version, stochastic DYNAMIC PROGRAMMING, has proven especially popular, being relatively easy to use in a wide variety of applications.

A stochastic dynamic programming model of optimal behaviour has the following components.

1 A dynamic state variable $X(t)$, generally multidimensional, representing the state of the organism and its environment at time t.

2 A decision set $D(X,t)$, depending on the current time and state, representing the set of feasible decisions, or actions.

3 A dynamic equation of state, $X(t+1) = G[X(t), d(t), w(t)]$ describing how the state variable changes over time in response to the decision $d(t)$ and stochastic environmental influences $w(t)$.

4 A specification of FITNESS, usually defined in terms of expected LIFETIME REPRODUCTIVE SUCCESS.

The algorithm used to calculate the optimal STRATEGY is known as dynamic programming. Dynamic programming models have been widely used to analyse behaviour and LIFE-HISTORY TRAITS that involve TRADE-OFFS between foraging benefits and predation risks. [C.W.C.]

optimal diet model One of the principal models of OPTIMAL FORAGING THEORY, the optimal diet model treats the decisions of a forager SEARCHING for food in a homogeneous environment. Several distinct types of food are encountered at random, each type being characterized by its relative abundance, its energy content and its HANDLING TIME (time required to consume the item). Upon encountering an item the forager may decide to eat it or to reject it: items having a low ratio of energy content to handling time may be rejected in the hope of encountering more profitable items. Based on the assumption that foragers will attempt to maximize the long-term average rate of energy gain, the standard optimal diet model predicts a THRESHOLD type of behaviour—items above the threshold are always accepted, those below rejected. Most published data, however, display partial preferences. [C.W.C.]

optimal foraging theory (OFT) Optimal foraging theory provides simple predictive models to explain why organisms prefer to eat certain items out of a set of available kinds of food. The classical models considered only maximization of intake rate in an unchanging world with complete information of the forager. Later models have explored other strategies, for example minimizing energy expenditure, risk minimization, avoiding predation. Likewise, stochastic dynamic foraging models have explored the effects on optimal foraging of changes in the environment, in the information an animal has about its environment, and in the physiological condition of the foraging animal.

Although net rate of energy gain is the 'currency' most commonly used as the basis for optimization, other currencies have been recognized as more appropriate to particular situations. The appropriateness of any currency may depend both on ecological circumstances and on the physiological state of the forager.

Optimal foraging theory has been criticized on the ground that animals cannot be expected to be optimal because of genetic CONSTRAINTS preventing animals from adapting fast enough to changes in the environment. However, protagonists of optimal foraging theory have argued that its aim is not to prove that animals are optimal but, rather, to make testable predictions about animal behaviour and so advance our knowledge about the great diversity of behaviours seen in nature. [J.V.A. & R.N.H.]

optimal group size Many species of animals live in groups, having either relatively permanent membership (as in family groups), or having a more ephemeral nature. The main advantages of grouping are thought to be: (i) increased protection from predators, resulting from combined VIGILANCE, or merely from safety in numbers; and (ii) increased foraging EFFICIENCY.

The FITNESS of an individual group member can be envisaged as a function of the size of the group, *N*. The optimal group size is then the size (N_1) at which individual fitness is maximized. [C.W.C.]

optimal harvest This idea involves a consideration of economic factors as well as the purely biological. The simple idea of a MAXIMUM SUSTAINABLE YIELD (MSY) is a purely biological concept. It may well be that the level of harvest which achieves, for example, maximum profit, may be greater or less than the MSY. Indeed, in certain situations it may be optimal, in the sense that profits are maximized, for a population to be driven to extinction. This occurs when the SUSTAINABLE YIELD is relatively small compared to the size of the stock. In such a situation, the yield is lower than that which can be obtained in market interest rates and hence it becomes attractive to overexploit the resource and turn it into money which can earn interest. [J.B.]

optimal reaction norms The optimal REACTION NORM for a TRAIT describes the optimal value of this trait for each environment in the range of environments encountered by the population. PHENOTYPIC PLASTICITY can be a mechanistic by-product of environmental influence. However, for a trait under STABILIZING SELECTION, the optimum will often change with the environment, and plasticity may thus be an adaptive response to environmental cues. The optimal reaction norm for such a trait describes the ideal adaptive plasticity: it is the reaction norm of a hypothetical GENOTYPE that in each environment produces the optimal PHENOTYPE. [T.J.K.]

optimal repair Allocation of metabolic resources to somatic repair (and maintenance) to maximize the FITNESS of an organism. This allocation is determined with regard to the TRADE-OFF that must be made between, on the one hand, protecting against age-associated increases in INTRINSIC MORTALITY by means of slowing the accumulation of somatic damage and, on the other hand, utilizing resources for growth and reproduction. *See* DISPOSABLE SOMA THEORY. [T.B.L.K.]

optimal reproductive effort The amount of effort that should be devoted to reproduction to maximize FITNESS. The two primary components are the age schedule of reproduction and the allocation of energy per reproductive episode. At one extreme the optimal REPRODUCTIVE EFFORT may be so high that only one breeding episode is possible (SEMELPARITY): such a situation is favoured by low adult survival and high rate of increase. ITEROPARITY, or repeated reproduction, is the more common situation. The optimal reproductive effort will be determined in large measure by the trade-offs between survival and reproductive effort, and between GROWTH and reproductive effort. The importance of the former is readily apparent; the importance of the latter resides in the fact that energy channelled into reproduction is not available for growth, and growth may be more strongly favoured because of the increased FECUNDITY or mating success that frequently accompanies an increase in SIZE. [D.A.R.]

optimal size at maturity AGE AND SIZE AT MATURITY are among the traits most closely related to the evolutionary success of individuals. Body SIZE is closely associated with FECUNDITY and survival in many organisms, so that there is SELECTION PRESSURE for large size. On the other hand, to become large, many organisms must delay their maturity, in particular if their GROWTH slows after they have matured. But delaying maturity decreases reproductive success because it decreases the likelihood that an individual will reach maturity, and increases its GENERATION TIME. Thus there is selection pressure for early maturity and therefore a small body size. Assuming that EVOLUTION will lead to a balance of these opposing selection pressures, life-history theory can predict the age and size at maturity that maximize reproductive success. [J.C.K.]

optimality models Optimality models are commonly used in evolutionary biology and in economics. They are based on a working assumption that the individual members of an ecological or economic COMMUNITY act as optimizers of some specified objective. In biological applications the optimality objective is usually taken to be Darwinian FITNESS, a measure related to survival and reproduction. [C.W.C.]

optimum pollution This concept stems from the recognition that it is often necessary to compromise between the benefits of POLLUTION control—in terms of human health and the environment—and the costs of the restrictions—in terms of lost social benefits to health, food production and general welfare. Thus many pesticides pollute; but they bring benefits to food production that may be crucial in the developing world.

The Royal Commission on Environmental Pollution, in a report published in 1972, describes the concept as 'unavoidably troublesome'. The Sixteenth Report (1992) comments:

> the concept of 'optimum' is 'coherent' (only) in the context of the analysis of a whole range of public decisions in which the harm caused by a certain amount of pollution needs to be compared with the harm which would be caused if *resources* were directed from other important purposes, not connected with pollution.

So optimality arguments are certainly difficult; objectively defining costs and benefits is intellectually challenging, and does not always lead to comfortable conclusions. But these trade-offs have to be kept in the public mind; it is important to realize that pollution control has to be paid for and that very often difficult decisions have to be made about the appropriateness of alternative products or courses of action. [P.C.]

ordination The arrangement of samples or sites along gradients on the basis of their species composition or environmental attributes. Ordination is the mathematical expression of the CONTINUUM concept in ecology.

Data for ordination typically consist of a table of values specifying the ABUNDANCE or presence of species in samples. Environmental data are usually also available. A convenient way to envisage the structure of the data is as two matrices, **A** and **B**, stacked one beside the other:

$$\mathbf{C} = [\mathbf{A}|\mathbf{B}] = [a_{ij}|b_{ik}]\,(i = 1, \ldots, m; j = 1, \ldots, n; k = 1, \ldots, p)$$

Rows of the matrix represent samples. The first block of n columns represents species; a_{ij} is the abundance of species j in sample i. The second block of p columns represents environmental variables; b_{ik} is the value of the kth environmental variable in sample i.

Ordination in two dimensions produces a diagram showing relations among samples and/or species. Samples that are near to one another in the ordination diagram are inferred to resemble one another in species composition and environmental attributes. There is a tacit assumption that samples with similar species have similar environments. [M.O.H.]

organic detritus Particulate material derived from dead organisms or from parts shed by living organisms. DETRITUS may be derived from the breakdown of land plants, benthic algae and grasses, PHYTOPLANKTON, and ZOOPLANKTON and zoobenthos. It provides a food source for certain organisms, referred to as detritivores. [V.F.]

organic loading The addition of abnormal amounts of dead organic matter to the environment—usually to a watercourse. [P.C.]

organic pollution Usually means contamination from wastes derived from living things that causes adverse effects on human health and/or ecological systems (*see also* SEWAGE POLLUTION). Can sometimes be intended as POLLUTION from organic chemicals. [P.C.]

organismic hypothesis According to Clements, *see* CHARACTERS IN ECOLOGY and CLIMAX):

> The unit of vegetation, the climax formation, is an organic entity. As an organism, the formation arises, grows, matures and dies. Its response to the habitat is shown in processes or functions and in structures that are the record as well as the result of these functions.
>
> Furthermore, each climax formation is able to reproduce itself, repeating with essential fidelity the stages of its development. The life-history of a formation is a complex but definite process, comparable in its chief features with the life-history of an individual plant.

This viewpoint was based on three specific ideas: (i) that species occur with characteristic companions in distinct communities; (ii) that there is at any site-type a closely repeatable SUCCESSION of species ('single pathway'); and (iii) that FACILITATION, as it is now called, is a general phenomenon in succession. These ideas have been largely displaced by the INDIVIDUALISTIC HYPOTHESIS, the concept of variable pathways, and the idea that many successions result from differences in life histories and tolerances without a major involvement of facilitation. The parallel between a plant 'formation' and an organism is generally rejected. [P.J.G.]

oscillations Regular fluctuations through a fixed cycle above and below some mean value. [V.F.]

osmoregulation The mechanisms whereby an organism maintains relatively constant water content and salt concentrations inside its body. [D.J.R.]

osmosis Diffusion of a solvent through a semipermeable membrane into a more concentrated solution. The osmotic pressure is the pressure that must be applied to oppose the osmotic flow. [J.G.]

outbreaks

1 Outbreaks are (small) EPIDEMICS of infectious disease localized in time and space. The term is usually used to describe an increase in MORBIDITY and MORTALITY due to an endemic infection. Outbreaks may often occur by chance.

2 Can also refer to PEST species.

[G.F.M.]

outwelling The outflow of NUTRIENTS from an ESTUARY or SALT MARSH system to shelf waters. [V.F.]

ova Female GAMETES, adapted for supporting early development. *See also* EGGS. [P.C.]

overcompensation The phenomenon occurring when density-dependent MORTALITY leads to POPULATION SIZE overshooting an equilibrium following disturbance. In the longer term, overcompensation is associated with an oscillatory approach to equilibrium, sustained cycles around equilibrium or CHAOS. Overcompensation is often the ecological consequence of SCRAMBLE COMPETITION. [R.H.S.]

overcrowding effect This term usually refers to DENSITY DEPENDENCE or INTRASPECIFIC COMPETITION, demonstrated by a reduction in the PER CAPITA rate of increase as POPULATION DENSITY increases. [R.H.S.]

overdispersion If a population is distributed randomly then samples from that population will follow a POISSON DISTRIBUTION and the MEAN and variance of the samples will be equal. If a population deviates from this distribution it is said to be either underdispersed—more ordered than random, giving a variance smaller than the mean of a SAMPLE—or overdispersed (or overdistributed)—more clumped than random, giving a variance greater than the mean. Therefore, overdispersion is, effectively, a synonym for AGGREGATION. The term is now used rather infrequently. [C.D.]

overfishing The practice of HARVESTING fish in quantities greater than can be replaced by RECRUITMENT to the population. [P.C.]

overgrazing GRAZING by domestic or wild animals when the food demand in the long term is greater than the primary productivity and the plant BIOMASS of the ecosystem goes into decline. [P.D.M.]

overgrowth competition COMPETITION for space on a two-dimensional surface shown, for example, by lichens and corals. One species (the competitive dominant) grows over another species and kills it. [B.S.]

overlapping generations The simultaneous occurrence in a single-species population of individuals of two or more different AGE CLASSES or life stages. [A.J.D. & T.J.K.]

overwintering A generalized term describing the way in which an organism passes through winter. Low TEMPERATURE is injurious and potentially lethal to most forms of life. Survival in climatic regions with a winter season is achieved by one of three main options: migration, hibernation, or freeze tolerance/supercooling. [J.S.B.]

oviparity (egg-laying) The production of a haploid GAMETE (an EGG) complete with sufficient NUTRIENTS to support the early GROWTH and development of the diploid offspring. DIPLOIDY is restored via fertilization by a sperm that is usually motile and contributes little or no nutrients to the embryo. One variation on this theme is for the meiosis of the egg to be triggered by fertilization. At least some development prior to hatching takes place after the egg has been laid, and hence external to the body of the female. Oviparity originally evolved as part of sexual reproduction, as an alternative to asexual forms of reproduction such as vegetative budding. It is present in virtually all phyla in the animal kingdom. It is not necessarily associated with sexual reproduction in all derived species; for example, asexual forms of reproduction such as PARTHENOGENESIS have evolved secondarily in many sexual lineages. [D.N.R.]

ovoviviparity (live-bearing) The production of a haploid GAMETE (an EGG) complete with sufficient NUTRIENTS to support the early GROWTH and development of diploid offspring. Eggs are internally fertilized, retained within the female throughout development, then young are born live. There is no further transfer of nutrients from mother to young after FERTILIZATION. [D.N.R.]

oxygen sag The fall and recovery in DISSOLVED OXYGEN downstream of a major source of organic effluent, such as a sewage works, brewery or paper mill. [P.C.]

ozone hole Thinning of stratospheric OZONE (O_3) over the Antarctic is a climatological feature, but since the late 1970s severe depletion has occurred each year in September and October, forming the ozone hole. Ozone depletion is most severe between altitudes of 16 and 22 km where the Antarctic lower STRATOSPHERE is extremely cold. On the surfaces of frozen cloud particles, chemical reactions take place which transfer chlorine from inactive to active species. Ozone is then catalytically destroyed, predominantly by the following cycle:

$$ClO + ClO + M \rightarrow Cl_2O_2 + M$$

$$Cl_2O_2 + h\nu + M \rightarrow 2Cl + O_2 + M$$

$$2Cl + 2O_3 \rightarrow 2ClO + 2O_2$$

where M is any molecule and $h\nu$ is a photon. The chlorine is largely produced from anthropogenic CFCs. [J.A.]

ozone layer Approximately 90% of atmospheric ozone (O_3) occurs in the ozone layer, in the STRATOSPHERE 10–50 km above the Earth's surface. Ozone absorbs ULTRAVIOLET RADIATION, thereby protecting the BIOSPHERE from its harmful effects. It is formed naturally by the action of sunlight on oxygen molecules. The sunlight also heats the stratosphere giving rise to an increase in TEMPERATURE with height. The temperature structure of the stratosphere determines atmospheric motions, which transport ozone from the region of high production in the tropics to high latitudes. [J.A.]

P

P : B ratio The rate of energy or matter accumulation per unit BIOMASS per unit time; i.e. the ratio of PRODUCTIVITY to biomass. Because biomass may increase over the period of measurement of net primary productivity the mean or average biomass for the period is used to calculate $P:B$. In early successional stages plant biomass and net primary production may increase in parallel; however, a point will be reached where further increase in biomass will not result in proportionate increases in net production, and may eventually lead to a decrease in net production. The same changes are seen at other trophic levels and with population growth, when food supply is limited. [S.P.L. & M.H.]

P : R ratio The ratio of PRODUCTION rates (P), which represent the 'benefits' of energy transfer in terms of formation of new somatic and reproductive tissues, to RESPIRATION (R), which represents the energy 'costs' of metabolism and activity. This is an ecological cost/benefit statistic indicating how efficiently a particular trophic unit is able to accumulate BIOMASS from the energy it assimilates from the TROPHIC LEVEL below. [M.H.]

palaeobotany The science of past plant life, as seen in the form of fossils preserved in sedimentary rocks. [W.G.C.]

palaeoclimatic reconstruction Palaeoclimatic reconstruction is normally based on proxy evidence for past CLIMATE CHANGE rather than on instrumental readings. Biological evidence for past climate change is contained mainly in the FOSSIL RECORD. Both macro- and microfossils of animals and plants have been used to track and quantify palaeoclimate change through time.

Fossil pollen data, in the form of pollen diagrams, are an important source of evidence of past vegetation distribution patterns which, to varying degrees, are under the control of climate. Several pollen records from south-eastern France, extending back to the penultimate glaciation (*c.* 180 000 BP), provide a record of climate change in terrestrial habitats situated in mid-northern latitudes. [M.O'C.]

palaeoecology Palaeoecology is the study of the ecology of past communities and past environments. It involves the reconstruction of the conditions and the flora and fauna of the past on the basis of a wide range of evidence, some biological, some geological. It is dependent upon the existence and survival of 'fossil' evidence of such past communities and conditions upon which the palaeoecologist can operate. [P.D.M.]

palaeolimnology The study of lake history based on the analysis of lake sediments. Evidence for environmental change, caused by POLLUTION or natural events, can be derived in particular from an analysis of the microfossil ASSEMBLAGES in the sediment. Of these the diatom, cladoceran and chironomid fossil records are the most important. Using modern calibration datasets, which relate biological distributions to environmental gradients, transfer functions can be generated to enable quantitative reconstructions of lakewater pH, phosphorus, SALINITY and TEMPERATURE to be made from the FOSSIL RECORD. [R.W.B.]

palsa mire A tundra mire type, whose name is derived from Finnish, found only within climates where the average air temperature remains at or below 0°C for more than 200 days in the year, roughly corresponding to the southerly limit of discontinuous permafrost. Palsa mires are often associated with northerly aapa mires, being found mainly on the flatter regions. The mire consists of elevated mounds, often 50 m or more in diameter and several metres in height, consisting of a core of frozen SILT and PEAT, with a peat and vegetation cover over their summits. These palsa hummocks are interspersed with pools of similar area, many of which become infilled by vegetation to form extensive wet lawns. [P.D.M.]

paludification The process by which soils become waterlogged and PEAT formation commences and/or spreads. [P.D.M.]

palynology Palynology is the study of pollen grains of flowering plants and gymnosperms and of spores from pteridophytes, bryophytes, algae and fungi. Such work can involve studies of structure, development, taxonomy, evolution, dispersal, preservation and STRATIGRAPHY. The applications of the studies are many, including allergies, forensic science and PALAEOECOLOGY. [P.D.M.]

pampas The extensive areas of TEMPERATE GRASSLANDS in South America. They are similar to the prairies of North America and the steppelands of Eurasia. Tall, medium and short grasses occur throughout the pampas, which dominate the Rio de la Plata lowlands in Argentina, Uruguay and Brazil. The most common grass genus is *Stipa* and clumps of trees occur in water-retaining depressions. [A.M.M.]

panbiogeography A biogeographic method that focuses on the spatiotemporal analysis of DISTRIBUTION patterns of organisms. It is distinct from phenetic BIOGEOGRAPHY, which investigates similarities between BIOTAS in terms of the number of taxa they have in common. The method involves drawing lines on a map, known as tracks, which connect the known distribution of related taxa in different areas. When two or more tracks of unrelated taxa coincide, they are combined into generalized tracks, which are thought to indicate the distribution of an ancestral biota before it vicariated. [A.H.]

panmictic Describing a population where mating occurs at random with respect to the distribution of GENOTYPES in the population. Panmictic populations are probably very rare in nature because spatial divisions in the environment result in inbreeding and preferences for males or females with certain PHENOTYPES result in ASSORTATIVE MATING. Totally random mating is one of the assumptions of the Hardy–Weinberg theorem. [D.A.B.]

paradox of enrichment The danger to a PREDATOR-limited population of victims that comes from increasing its supply of the very RESOURCES whose shortage is killing victim individuals or reducing their birth rates. The increase in resources can sometimes destabilize the interaction with predators and lead to the extinction of the entire victim population. [M.L.R.]

parallel bottom community The similarity of species composition in temperate and arctic bottom communities of widely separated locations. [V.F.]

parallel evolution The occurrence of homoplasy (independently evolved traits) in paraphyletic taxa. An older, more restrictive use of the term is for homoplasy derived from the same plesiomorphic background. [D.R.B.]

parameters The summary measurements or characteristics of a POPULATION. These are represented by Greek letters, for example, the MEAN (μ) and the variance (σ^2). [B.S.]

parametric statistics Statistical tests are frequently classified into parametric and non-parametric tests. Parametric tests are older and require quantitative information/observations to follow a NORMAL DISTRIBUTION and to have similar VARIABILITY (homogeneous variances). Parametric tests cannot be carried out on qualitative information, i.e. ranked information or information in categories. If the underlying conditions are met then parametric tests are usually more powerful (*see* STATISTICAL POWER). Examples of parametric tests/techniques are *t* TEST, PRODUCT–MOMENT CORRELATION COEFFICIENT and ANALYSIS OF VARIANCE. *Cf.* NON-PARAMETRIC STATISTICS. [B.S.]

páramo Humid arctic/alpine meadows and SCRUB, occurring in the Andes and kept moist by mist. [P.C.]

parapatric speciation Though there is potential GENE FLOW between populations which are parapatric, sufficient isolation can occur for SPECIATION if gene flow is low and the difference in SELECTION PRESSURES between semi-isolated groups is high. [P.C.]

paraphyletic group An incomplete evolutionary unit in which one or more descendants of a particular ancestor have been excluded from the group. For example, a group that placed all cartilaginous fish, ray-finned fish, coelacanths and lungfish together in one TAXON (called 'fish') would be paraphyletic because lungfish and coelacanths share a common ancestor with tetrapod vertebrates that is not also shared with cartilaginous and ray-finned fish. [D.R.B.]

parasite An organism that interacts with another organism (the HOST) to reduce the FITNESS (survival and/or reproduction) of the host, and increase the fitness of the parasite, such that the parasite does not have to kill the host in order to derive the benefit (although the host may die as a result of the interaction). The term parasite is traditionally used for macroparasites, although increasingly it is used to include microparasites. *See also* PARASITISM; SPECIES INTERACTIONS. [G.F.M.]

parasitism An intimate and usually obligatory relationship between two organisms in which, essentially, one organism (the PARASITE) is exploiting RESOURCES of the other organism (the HOST) to the latter's disadvantage. In general, parasitism involves a long-lasting association between the two parties. However, it is difficult to give a universal definition, as parasitism is a widespread and extremely diverse phenomenon. Parasitism is different from predation (where the prey has no lasting association with the PREDATOR), from COMPETITION (where both parties may suffer), and from COMMENSALISM (where the negative effects are absent or negligible). [P.S.H.]

parasitoid Any insect whose larvae develop by FEEDING on the bodies of other arthropods, usually insects. It is estimated that 0.8–2 million species of insects are parasitoids. The parasitoid LARVA or larvae eventually kill their HOST, and a single host provides all the food required for development into an adult. [H.C.J.G.]

parent–offspring conflict There is said to be parent–offspring conflict if a trait that increases the reproductive success of an individual offspring can only do so at the expense of the reproductive success of its parent(s) and vice versa. In short, parents and their offspring need not have precisely the same set of interests. When their interests are not identical there is said to be a potential conflict between the two. [L.D.H.]

parental care Any form of behaviour by an adult which benefits its offspring. [J.D.R.]

parental investment Any behaviour by a parent which increases the FITNESS of that parent's offspring at a cost to the parent's own future reproductive success, including activities such as care of offspring or the provisioning of EGGS. Parental investment should be measured in fitness benefit to the offspring, but the term is often used loosely to mean energy invested in PARENTAL CARE. [J.C.D.]

parental manipulation hypothesis The hypothesis that the EVOLUTION of eusociality in animals is based on the retention of mature offspring due to manipulation by their parents. Parental manipulation of offspring has been involved in the evolution of eusociality and in sociality in general because offspring forego reproduction by staying with their parents and providing help for their parents' reproduction rather than reproducing on their own. Other factors such as MUTUALISM and KIN SELECTION also appear to have played a role in the evolution of eusociality. [A.P.M.]

Park's experiments In a series of studies in the 1950s and 1960s, Thomas Park carried out 'bottle' experiments on two species of flour beetle, *Tribolium confusum* and *T. castaneum*. Both species are frequently found as stored-product pests and they can be conveniently kept under laboratory conditions. Populations were kept in glass tubes with 8 g of medium (95% flour + 5% yeast). All adults, pupae and larvae were counted every 30 days, and both single-species and mixed-species populations were kept. An interesting aspect of the EXPERIMENTAL DESIGN, which proved instructive, was that populations were kept at a number of temperatures (24°C, 29°C and 34°C) and relative humidities (30% and 70%), and each type of population (single or mixed), under all six environmental conditions, was replicated 30 times.

Park found that while single-species populations generally persisted, mixed-species populations did not. In each population only one species persisted, the other was eliminated. However, unlike most previous 'bottle' experiments (*see* GAUSE'S EXPERIMENTS) it was not always the same species that was eliminated, even under the same environmental conditions. This 'indeterminate' outcome of COMPETITION is, however, one of the predicted possible outcomes of the LOTKA–VOLTERRA MODEL. [B.S.]

parsimony There is an infinite number of hypotheses that might be suggested by a particular set of data. Scientists and philosophers rely on the principle of simplicity (parsimony) to pick the simplest hypothesis; the hypothesis that explains the data in the most economical manner. Simplicity, economy and parsimony all refer to the same general concept. However, the nature of this concept is not well understood and has been debated by philosophers without resolution for many years. [D.R.B.]

parthenogenesis The production of a viable embryo from an unfertilized EGG. This may occur sporadically (ptychoparthenogenesis) or as a permanent feature of the LIFE CYCLE. Having many independent origins in higher plants and animals, parthenogenesis occurs through a wide variety of cytogenetic mechanisms. Parthenogenetic eggs may be haploid, as a result of normal oogenesis, or diploid owing to aberrant oogenesis. [R.N.H.]

partial refuge A REFUGE that is only available for part of the time or for a proportion of the population. [B.S.]

particle-size selectivity The ingestion of a size range of particles (usually of PHYTOPLANKTON, or SEDIMENT particles) in a proportion different from that available in the surrounding medium. SUSPENSION-FEEDING organisms that feed on phytoplankton may selectively ingest larger-sized species because food value is often proportional to phytoplankton cell size. Deposit-feeding organisms often selectively ingest smaller sediment particles because these have a larger surface area for the attachment of bacteria or nutritional organic molecules, and can therefore have a higher food value per unit mass than larger sediment particles. Particle-size selectivity can occur via behavioural mechanisms or as a result of fixed properties of the feeding appendages. [V.F.]

particulate organic matter (POM) Dead organic material (DETRITUS) either suspended in the water column or settled on the substratum of aquatic habitats. It may be divided into COARSE-PARTICULATE ORGANIC MATTER (CPOM), such as leaf fragments, and FINE-PARTICULATE ORGANIC MATTER (FPOM), such as sediments. [P.C.]

passive dispersal A process that leads to the movement of individuals away from others with which they had been associated (often parents or siblings) without any locomotory effort by the individuals concerned. In contrast with ACTIVE DISPERSAL, the individual is carried passively, for example by wind or water. [R.H.S.]

passive suspension feeding The trapping of suspended particles by animals that do not propel water more than a few particle diameters from their collection appendages and that appear to trap particles by direct interception and by positioning themselves in BOUNDARY LAYER situations where particle fluxes are high. In contrast, active suspension-feeders move water to create their own feeding currents. [V.F.]

pasture GRAZING land that is managed to provide appropriate YIELD in terms of quantity and sometimes quality of herbs and grasses. [P.C.]

patch dynamics An aspect of metapopulation dynamics that describes the proportion $p(t)$ of HABITAT patches that is occupied at time t. Two parameters describe the key processes: the rate of colonization of empty patches (m); and the rate of extinction of local populations within patches (e). [R.H.S.]

patch use It is assumed that foragers perceive food as occurring in discrete patches separated by unproductive habitat. Two important behavioural decisions considered by OPTIMAL FORAGING THEORY (OFT) are which patches to exploit and how long to remain foraging in a patch before moving to another. The decision of which patches to exploit has been studied in relation to risk, of starving, of being outcompeted or of being killed by predators while foraging.

The decision of how long to stay in a patch has most successfully been addressed by a particular model, the marginal value theorem (MVT), which has emerged as one of the most enduring and influential models in BEHAVIOURAL ECOLOGY. The MVT uses the energy maximization premise (EMP) to predict optimal patch-residence time on the basis of relative productivity of the exploited patch and on distances between patches.

A different approach is required to model 'central-place foragers', such as birds provisioning a nest of young, limpets making foraging excursions from a refuge, or deposit-feeding invertebrates centred upon a burrow. For provisioners, behavioural decisions include how much food to carry in a single load back to the central place, how much time to spend foraging at successive distances from the central place, and how selectively to forage at these distances. It is predicted that food load, foraging time and selectivity should decrease further away from the central place, and that these trends should be reinforced by increasing productivity of the habitat and decreasing energy demands on the provisioner. Predictions for grazers and deposit-feeders foraging from a refuge or burrow include the optimal apportionment of time between foraging or refuging and the degree to which overlap should be avoided in successive foraging excursions. [R.N.H.]

patchy habitat An environment in which organisms tend to aggregate in patches where a resource (such as food or a breeding site) is at high density. The areas between patches may have no usable resource at all. A patchy HABITAT is defined in relation to a particular species. A particular stage in the life cycle of a species (usually juveniles) may be confined to a patch. Examples of patchy habitats include seeds (infested by weevils), fruits (infested by fruit flies), carcasses (infested by blowflies) and puddles (infested by mosquitoes). [R.H.S.]

paternity Male parent of offspring. The assessment of paternity of broods is made by genetic methods such as external polymorphisms, enzyme POLYMORPHISM or GENETIC FINGERPRINTING including microsatellite methods.

Extra-pair paternity. The occurrence of extra-pair paternity (offspring fathered by males other than the male socially bonded with a female) has been studied in a large number of species during recent years. It varies from complete certainty of paternity in some species of birds, to species in which more than two-thirds of all offspring are fathered by males other than the resident male.

Control of paternity. The control of paternity appears generally to be determined by females because females rather than males determine the fate of sperm within the female reproductive tract. Females also often control which male is able to copulate shortly before the time of fertilization.

Male assessment of paternity. Pair males appear to be able to assess their certainty of paternity to some extent. For example, males of some species engage in copulations with their mates relative to the sexual interest that neighbouring males have shown for the female. Pair males also provide PARENTAL CARE for offspring relative to their certainty of paternity. [A.P.M.]

path analysis Path analysis was introduced by Sewall Wright. It resembles multiple regression as a technique for disentangling CORRELATION and causation, but is more general, allowing tests of alternative models of causation. It is also used to predict correlation when causal pathways are postulated, and for purposes of graphical illustration. [T.D.P.]

pathogen A medical (human and veterinary) term for PARASITE, especially one associated with a high degree of MORBIDITY or MORTALITY. [G.F.M.]

pattern analysis A procedure used to detect the departure from randomness of the spatial arrangement of plants within an area (pattern). Distributions may be random, regular or, most commonly, clumped (*see* REGULARITY). [A.J.W.]

peat An unconsolidated, stratified organic material derived largely from undecomposed or partially decomposed vegetation. [P.D.M.]

peatlands Areas occupied by peat-forming vegetation (*see* MIRE; PEAT). Such areas are at least partially waterlogged and the rate of microbial DECOMPOSITION fails to keep up with that of plant litter formation, hence peat accumulates. [P.D.M.]

pedogenesis The myriad of processes leading to the formation of soils. [E.A.F.]

pedology The study of SOIL in all its aspects. [S.R.J.W.]

pedosphere The upper part of the Earth's CRUST that is occupied by soils. [E.A.F.]

pelagic Living in the water column (for marine organisms), seaward of the shelf-slope break. Pelagic organisms are adapted to maintain their position in the water column by having a shape (i.e. high surface area) or body composition (i.e. rich in lipid) that enhances buoyancy or by being able to swim actively. Some organisms are only pelagic for a part of their life cycle (meroplankton), whereas others are pelagic for their entire life cycle (holoplankton). Many pelagic organisms migrate vertically in the water column, sometimes covering great distances, on a daily or seasonal basis in order to meet energy requirements or escape predators. [V.F.]

per capita Latin: literally, according to heads. This term is used to express a quantity as a function of an individual. [R.C.]

percentile Any of the 99 actual or notional points that divide a DISTRIBUTION or set of observations into 100 equal parts. [B.S.]

perennial A plant (sometimes animal) living for a number of years. [P.O.]

peripatric speciation SPECIATION by evolution of ISOLATING MECHANISMS in populations located at the periphery of a species range. It is a special case of ALLOPATRIC SPECIATION, and is based on the observation that marginal, isolated populations often differ markedly from the main-range populations. [J.M.Sz.]

periphyton A term sometimes applied to the community of bacteria and algae that occurs in the littoral or benthic zones of LAKES, and found attached to natural substrates such as larger plants (epiphyton), stones and rocks (epilithon), and sand grains (epipsammon). It also includes communities of motile algae that live on and beneath mud surfaces (the epipelon). [R.W.B.]

permanence A concept of global community stability (*see* STABILITY, COMMUNITIES). Whereas LOCAL STABILITY measures examine whether a community tends to return to its original state following perturbation, permanence simply examines whether populations tend to coexist, irrespective of the stability of the equilibrium point. This distinction is important, since proving that an equilibrium is unstable does not necessarily imply that species cannot coexist; species may coexist, for example, on periodic or chaotic attractors. [R.P.F.]

pest An animal (or, occasionally, a plant) that consumes and/or damages living or dead plant materials (less usually animal materials) intended for human consumption or use; for example, animals that consume CROPS, animal feeds or stored products. Such pests are effectively competitors with humans. [P.C.]

pest control Attempts to eradicate or reduce densities of pests to levels where they are no longer effective competitors with humans. Control can be by chemical means (PESTICIDE) or biological means (*see* BIOLOGICAL CONTROL), or a combination of these (*see* INTEGRATED PEST MANAGEMENT). [P.C.]

pesticide A substance that kills pests. Pesticides can be specific or non-specific. *See also* BIOCIDE. [P.C.]

pesticide treadmill The seemingly endless search for new PESTICIDES that is required to counter the development of pest RESISTANCE to existing pesticides. [P.C.]

pH A logarithmic scale indicating the activity (or, approximately, the concentration) of hydrogen ions in solution:

$$pH = -\log_{10}[H^+]$$

where $[H^+]$ is the concentration of H^+ in moles per litre.

The pH of rain in clean environments is about 5, due to the carbonic acid/carbon dioxide equilibrium:

$$CO_2 + H_2O \leftrightarrow H_2CO_3 \leftrightarrow H^+ + HCO_3^-$$

In polluted environments it may be much lower, about 4; this 'acid rain' contains sulphuric and nitric acids formed from the atmospheric oxidation of nitrogen and sulphur oxides. The pH of surface and ground waters is very variable, ranging from <4 to >8, and that of soils even more so (2 to >10). [K.A.S.]

phenetic species concept A species concept based on phenetics, or NUMERICAL TAXONOMY. A set of organisms is classified as a species when its overall phenotypic similarity (and/or its phenetic distance to another set of organisms) exceeds a certain threshold. The phenetic species concept attributes no special significance to whether or not individuals of a species interbreed or share a common ecological NICHE. [W.W.W.]

phenology The study of periodic biological phenomena, especially in relation to seasonal environmental cycles. Phenomena typically studied in this regard include life-history features (GROWTH, reproduction, death), migration and DORMANCY (hibernation, DIAPAUSE). [R.B.H.]

phenotype Any structural or functional feature (i.e. TRAIT) of an organism other than genetic information. This definition encompasses physical and chemical characters of the organism (biochemistry, physiology, morphology), its behaviour and the life history. The 'extended phenotype' is a broader category, including an individual's artefacts, as well as its influence on other organisms and the environment. The phenotype is a product of the interaction between the GENOTYPE and the environment. [T.J.K.]

phenotypic plasticity The capacity of organisms for morphological, physiological or life-history modifications arising from either strictly environmental influences or GENOTYPE*ENVIRONMENT INTERACTIONS. Such modifications are usefully partitioned into two categories.

Phenotypic modulation results from exposure of developing organisms to different environmental conditions and is manifested as a CONTINUUM of PHENOTYPES. The capacity for this kind of modulation may be heritable and thus subject to NATURAL SELECTION, or it may be an unselected consequence of developmental or physiological CONSTRAINTS in a continuously variable environment. The outcome of phenotypic modulation is known as the REACTION NORM, and has been found to vary among individuals in both plants and animals with different genotypes.

Developmental conversion (also termed conditional choice or developmental switch) results from an environmentally induced switch between alternative developmental programmes, resulting in discontinuous phenotypes (usually two). Such a switch is often associated with discrete differences in the environment and is of apparent adaptive value. Typically, the switch is made in the juvenile stage in response to a specific environmental cue, committing the adult to an irreversible morphological type. [C.M.L.]

pheromone A chemical compound emitted by one organism which alters the behaviour or physiology of another member of the same species. [R.H.C.]

philopatry Tendency of an animal to return to, or stay in, its home area. [P.C.]

phoretic association Phoresy is a method of DISPERSAL whereby a species is transported by clinging to the body of another species and then drops off after a period of time. Phoretic association is a term which describes this relationship between two species. [S.J.H.]

phosphorus cycle The movement of phosphorus (P) between the BIOSPHERE, LITHOSPHERE and HYDROSPHERE, and the associated transformations between its different chemical forms, from the release of P from its main reservoir in apatites (complex calcium phosphates) and other minerals to the ultimate formation of new minerals. Weathering converts some P into exchangeable phosphate anions, which are absorbed by plants and pass into the biological cycle. The uptake is often promoted by the association between roots and mycorrhizal fungi. P is present in all living organisms, for example in DNA and phospholipids. Organic residues returned to the SOIL (and aquatic sediments) are mineralized by microorganisms, releasing phosphate again for return to the biological cycle or precipitation as insoluble phosphates of iron, aluminium or manganese (acid conditions) or of calcium (alkaline conditions). P is often the limiting nutrient in aquatic ecosystems, and increased supply can result in ALGAL BLOOMS and other problems. [K.A.S.]

photic zone The uppermost zone of a water body in which the penetration of light is sufficient for PHOTOSYNTHESIS to exceed RESPIRATION. [J.L.]

photoperiod The length of the day (*see* DAYLENGTH); or the length of the light period in experimental studies. [J.G.]

photosynthesis Photosynthesis is the process in which light energy is used to reduce carbon dioxide (CO_2) to sugars and other carbohydrates and in which molecular oxygen (O_2) is evolved. It is central to all life, in that photosynthesis ultimately produces all sources of food. [J.I.L.M.]

photosynthetic capacity The maximum rate of carbon dioxide (CO_2) uptake by plants under optimal conditions in the natural environment. [J.G.]

phototaxis Orientation or movement in response to light. [V.F.]

phototrophic Describing organisms that utilize light as an energy source. [V.F.]

phyletic gradualism A model of EVOLUTION in which species are presumed to diverge gradually after SPECIATION. [V.F.]

phylogenetic systematics Phylogenetic systematics is the name used by Willi Hennig, a German specialist in dipteran (fly) systematics, for an approach to producing what he termed a 'general reference system for comparative biology'. Hennig contrasted special interest reference systems (CLASSIFICATIONS) with a general reference system. He reasoned that there are many possible criteria for constructing special interest classifications, any one of which could produce a classification that would be maximally efficient for a particular purpose, but would have limited applications. A general reference system, by contrast, would provide the maximum amount of utility to the maximum number of biologists interested in a wide variety of topics.

Hennig reasoned that if biological EVOLUTION was responsible for producing the biological DIVERSITY around us, then the logical criterion for the general reference system would be one that classified species according to their evolutionary relationships. He proposed a logical method by which biologists could use similarities among species as evidence of relationships. [D.R.B.]

phylogeny The history of life. [D.R.B.]

physical factor Any non-living, non-chemical factor that influences biological processes. Physical factors include: heat, light, electricity, magnetism, radiation, gravity and pressure. *Cf.* BIOTIC FACTORS. [P.C.]

physiognomy Form or appearance. The physiognomy of a plant can be used to measure indirectly the ENVIRONMENT because it represents a suite of ADAPTATIONS to that environment. The physiognomic method was commonly used to characterize different vegetations but has been largely replaced by the phytosociological method, which uses the associations of different species. Nevertheless, physiognomy is still useful in characterizing the environment of ancient habitats from the physiognomies of plant macrofossil ASSEMBLAGES. For example, well-defined growth rings are found in the wood tissue of trees which have experienced a seasonal climate. [M.I.]

physiological ecology Physiological ECOLOGY has traditionally been concerned with explaining distributions in terms of the interaction between physiological processes and physical and chemical conditions; for example, how TEMPERATURE influences RESPIRATION; how chemical quality of the environment influences the *internal milieu* of organisms; explaining the EVOLUTION of physiological ADAPTATIONS in terms of environmental conditions. Exploring these issues has relied upon a comparative approach that has analysed divergences between related organisms in different environments (e.g. contrasting physiologies of organisms in different zones on the seashore) or the convergence between unrelated organisms in similar environments (e.g. the evolution of ANAEROBIC METABOLISM in endoparasites).

Recently, physiological ecology has become more concerned with issues of POPULATION DYNAMICS:

explaining how physiological processes influence POPULATION SIZE changes through time; i.e. by influencing growth rates and hence time to breeding, reproductive output and survival probability; explaining how different physiologies and the dynamics that are associated with them are favoured under different ecological circumstances so leading to physiological and life-cycle adaptations. Exploring these issues has relied on an interaction between the development of theoretical models and the design of experiments to test the hypotheses incorporated into the models and the predictions that emerge from them. [P.C.]

physiological time Physical TIME is invariant. However, whereas 1 minute is a long time with respect to the physiological processes and total LIFESPAN of a bacterium, it is a short time in the physiological processes and total lifespan of an elephant. Physiological time probably scales with body mass in the same way as metabolic rate. [P.C.]

physiological trade-off TRADE-OFFS are important in considering microevolutionary processes, especially in terms of life-cycle/life-history evolution (*see* LIFE-HISTORY EVOLUTION). If there were no trade-offs one would expect NATURAL SELECTION to maximize growth, reproduction and survival simultaneously; that it does not is powerful evidence for trade-offs even though their experimental demonstration has been elusive.

Physiological trade-offs are themselves based on the laws of thermodynamics, which specify that energy can neither be destroyed nor created, just transformed, and that these transformations lead to more entropy, which is usually manifest as heat loss. Thus if the organism is represented as a system that takes in energy (and other RESOURCES) as food and then allocates it between physiological demands such as maintenance, growth and reproduction, then the laws of thermodynamics require that if more energy is allocated to one process, less will be available for another. This is the basis of physiological trade-offs, sometimes known as the principle of allocation. It is used to explain trade-offs between life-cycle traits such as growth and survival (related to maintenance), growth and reproduction, and reproduction and survival. Detailed knowledge of the physiological processes involved may allow precise functional relationships to be established between physiological and life-cycle traits that are thought to be trade-offs, and hence to construct the trade-off curves that are so important in the rigorous development of life-cycle theory.

Various complications may blur these physiological trade-offs. For example: the input of resource (in FEEDING and PHOTOSYNTHESIS) may not be fixed but variable; body stores may be used to supplement the supply of resources to metabolism; some resource other than energy (e.g. phosphorus, trace elements) may be limiting; and the genetic basis of physiological trade-offs may be complicated by environmental effects. [P.C.]

phytoalexins A series of chemical substances produced by plants, often in response to wounding, that have antifungal properties. [P.D.M.]

phytochrome A protein pigment present in trace quantities in the leaves, stems and seeds of green plants, which changes its form depending on the ratio of red to far-red light. It is involved in the detection of: (i) shadelight; (ii) DAYLENGTH; and (iii) depth of the seed in the soil, and enables the plant to respond to the presence of competitors, and to adjust its LIFE CYCLE to the seasons. [J.G.]

phytogeography The study of the geographical DISTRIBUTION of plants on the Earth's surface. [A.M.M.]

phytophagous FEEDING on plants. [G.A.F.H.]

phytoplankton The collective name for groups of algae and bacteria that are adapted to living suspended in the open waters of marine and freshwater environments. [R.W.B.]

phytosociology The CLASSIFICATION of VEGETATION into units based upon floristic analysis. [P.D.M.]

phytotelmata Phytotelmata are small water bodies that occur in or on plants. Good examples are water-filled tree holes and the water contained in pitcher plants. [S.J.H.]

picoplankton PLANKTON, mostly bacteria, measuring 0.2–2.0 μm in size. [V.F.]

pioneer species Early-successional FOREST-tree species. [M.J.C.]

plankton The term plankton refers to very small, mainly microscopic organisms—bacteria (bacterioplankton), algae and cyanobacteria (PHYTOPLANKTON) or animals (ZOOPLANKTON)—that live suspended in the open water of freshwater and marine environments. [R.W.B.]

plant association An ASSEMBLAGE of plants that frequently occurs in combination in nature—a plant COMMUNITY. [P.D.M.]

plant defence theory Plants as AUTOTROPHS provide an almost complete nutritional base for numerous animals, and for pathogenic and saprophytic fungi and bacteria. Plants synthesize and accumulate a range of readily digestible carbohydrates, fats and proteins, and provide the majority of vitamins and essential minerals. The basis for plant defence theory is the assumption that, without defences, plants would be eaten out of existence by animals or would be eliminated by ubiquitous pathogenic microbes. Support for a defence theory comes from the observation that some plant species are subjected to heavy GRAZING (e.g. pasture grasses) while others consistently seem to escape predation (e.g. ragworts, *Senecio* spp.).

Defences may be structural or chemical. Structural defences include tough epidermis, thick bark, spines—in which the whole leaf is modified for defence (taken to extreme in the Cactaceae)—or prickles—which typically emerge from the leaf or stem epidermis, a feature of the Rosaceae. Several thousand compounds with some defensive function have been described, including alkaloids, phenols and quinones, terpenoids and steroids (*see* PLANT TOXINS). [G.A.F.H.]

plant–herbivore interactions Like all biological, especially trophic, interactions the relationships between animal feeder and plant food can lead to responses in each or both on an ecological (short) time-scale and on an evolutionary (long) time-scale. Because the land is generally green, there has been a presumption that herbivores are not food limited, but instead are controlled by organisms that feed on them. This is more generally true for terrestrial than aquatic systems; in the former the tissues necessary for providing support in air (CELLULOSE/lignin) are effective in preventing animal GRAZING and digestion. So more plant PRODUCTION goes to litter and soil on land than to sediment in waters. Nevertheless, terrestrial herbivores have evolved methods for ingesting plant materials (e.g. grinding teeth; or sharp suctorial mouthparts for removing sap) and can have local effects on survival, growth, reproduction and hence density of plant cover. In turn, plants (both terrestrial and aquatic) have evolved defence systems including physical spines and chemical TOXINS.

It is also of considerable importance that herbivory of reproductive tissues, or materials produced by them, can be of mutual benefit. Herbivores that consume pollen and nectar also transfer pollen from plant to plant; and fruit-feeders may disperse seeds. [P.C.]

plant toxins Toxic compounds are widespread throughout most plant groups, including angiosperms, gymnosperms, pteridophytes and bryophytes. Of the 10 000 or so characterized terpenoids, 8000 phenolic compounds and several thousand alkaloids, probably the majority induce toxic responses, at least *in vitro*. Whether this toxic effect is the natural function has only been confirmed for a few hundred compounds. Even fewer TOXINS are known, with certainty, to confer a significant ecological benefit to the producer. However, intuitively and logically it is most likely that the majority of plant toxins function in the control of herbivory—ranging from the GRAZING of large mammals and larvae of moths to the FEEDING of aphids. Plant toxins also act as a potent defence against fungal, bacterial and viral challenge. [G.A.F.H.]

plate tectonics The process by which lithospheric plates are moved across the Earth's surface to collide (i.e. at subduction zones), slide by one another (i.e. to form transform faults), or move away from each other (i.e. at divergence zones) to produce the topographic configuration of the Earth. Plate movements are driven by heat (arising from radioactive decay) escaping from the Earth's core and mantle, which causes convective movements within the mantle. Plates move away from each other along midocean ridges, and new CRUST forms at the ridges as magma erupts through fissures in the floor. As the new crust cools, it becomes denser and sinks. Crust is destroyed at the ocean trenches as gravity draws one plate under another and the old, dense LITHOSPHERE is subducted. [V.F.]

playas Shallow, circular basins of ALLUVIAL fans in plains and DESERT landscapes, they are also known as bajadas. Fine-textured SOIL and accumulation of salts make playa lakes a poorly drained habitat; precipitation provides ephemeral water inflow, with each of the playas and its WATERSHED developing vegetation in a CLOSED SYSTEM. [S.K.J.]

pleiotropy The genetic phenomenon whereby a single allelic change has effects on multiple characters, rather than on one character only. [M.R.R.]

Podzol A description of soils. The Podzol concept was developed in Russia, where the principal criterion is the presence of a leached horizon, as indicated by the origin of the term from the Russian words *pod* under, and *zola* ash; translated it means 'ashes under'—when these soils are cultivated, the plough turns up the leached horizon, which resembles ashes. Therefore all soils with leached horizons are Podzols, including Acrisols (Ultisols) and Luvisols (Alfisols). [E.A.F.]

poikilotherm An organism with a highly variable body TEMPERATURE. In contrast, a HOMEOTHERM has a relatively constant body temperature. [R.B.H.]

point quadrat A frame, often wooden, housing a row of pins (frequently 10), which are lowered vertically on to VEGETATION, the points of contact giving an objective measure of COVER. The pins resemble knitting needles, and are best guided through two laths to prevent parallax effects. The tip diameter is as small as possible (tip size correlates with exaggeration of cover values). A succession of measurements (e.g. 10 positions of the pin-frame along a line) are taken. [A.J.W.]

point source An identifiable, discrete origin of contamination or POLLUTION, such as the outlet of a pipe or chimney stack. [P.C.]

Poisson distribution A discrete PROBABILITY DISTRIBUTION that describes situations where rare observations (e.g. individuals in a sampling unit) follow a RANDOM DISTRIBUTION in either space or time. [B.S.]

pollard A tree whose top has been cut off well above the ground and which has sprouted again from the stump. [G.F.P.]

pollen analysis The extraction, identification and quantitative assessment of fossil pollen grains in stratified lake sediments or PEAT deposits as a means of reconstructing the past history of vegetation in a region. *See also* PALYNOLOGY. [P.D.M.]

pollination Pollination is the process of transferring pollen from anthers to receptive stigma in the flowers of seed plants. It is, in essence, sexual intercourse by post. A pollen grain contains a sperm cell within a DESICCATION-resistant coat. In order to achieve FERTILIZATION, it must be deposited on the stigmatic surface, germinate, and send a long, narrow pollen tube through the tissue of the style to the recipient's ovary. The nucleus of the sperm cell must then pass down this tube to fertilize an ovule (containing an egg cell) in order to produce a fertile seed.

The process is complicated because most plants, being SESSILE, must depend on some intermediary (whether biotic or abiotic) to transfer their GAMETES to potential sexual partners. The variety of processes by which this is accomplished makes pollination biology an area rich in natural history, and creates an unusually fertile testing ground for theories in ECOLOGY and evolutionary biology. [W.E.K. & J.Cr.]

pollution The release of a by-product of human activity—chemical or physical—that causes HARM to human health and/or the natural environment; contamination causing adverse effects. [P.C.]

polyandry An unusual MATING SYSTEM, found in birds and mammals, in which some females in a population mate with multiple males. [J.C.D.]

polygamy A MATING SYSTEM in which (i) some of the males in a population mate with more than one female during a single breeding season (POLYGYNY); or (ii) some of the females mate with more than one male (POLYANDRY); or (iii) some members of each sex have multiple partners (polygynandry, or PROMISCUITY); the opposite of MONOGAMY. [J.C.D.]

polygenic trait A TRAIT for which genetic variation between individuals is caused by the segregation of ALLELES at multiple loci (QUANTITATIVE TRAIT loci, or polygenes). [T.F.C.M.]

polygyny A MATING SYSTEM in which some of the males in a population mate with more than one female during a breeding season. [J.C.D.]

polymorphism The occurrence of discrete PHENOTYPES or forms within a population(s) of a species. It can refer to: a LIFE CYCLE (e.g. medusae/polyps of cnidarians); individuals with different GENOTYPES in a population (e.g. melanic and non-melanic moths); individuals with different functions within a COLONY/social group (e.g. feeding and reproductive individuals in colonial hydroids; workers and queens in social insects). [P.M.B. & P.C.]

polyphagous Describing organisms that are able to exploit a wide range of food RESOURCES. [R.C.]

polyploidy The condition of possessing more than two entire chromosome complements. Most eukaryotic organisms are diploid (i.e. have two chromosome sets) in the majority of their cells but produce haploid GAMETES (with one set of chromosomes). Polyploidy can arise by the addition of one or more extra sets of chromosomes, identical to the normal haploid complement of the same species (i.e. autopolyploidy) or by the fusion of genomes of two different species, for example by hybridization (i.e. allopolyploidy). Polyploidy can occur, for example, when a diploid ovum arises, as a result of suppression of meiosis, and is fertilized by a normal haploid sperm, or when a normal haploid ovum is fertilized by two sperm. The resulting offspring will have three of each kind of chromosome, and is called a triploid. An organism with four of each kind of chromosome is called a tetraploid. Polyploidy involving an even number of sets of chromosomes is more common than that resulting in an odd number of sets of chromosomes.

Polyploids are often reproductively isolated from their diploid ancestors, and particularly in plants (in which the polyploid descendants can also reproduce asexually) polyploidy has played an important role in the EVOLUTION of new species. Polyploidy is rare in most animal groups, presumably because it tends to interfere fatally with meiosis by generating chromosome imbalance, but is well known in lizards, amphibians and fish. In addition to arising in germinal tissue, polyploidy can also occur in somatic tissue (e.g. in the mammalian liver), probably as a result of abnormal cell division cycles. [V.F.]

polyterritoriality The simultaneous possession of two or more spatially separate breeding territories by FACULTATIVE polygynists (*see* POLYGYNY). [A.P.M.]

pools, in streams *See* WITHIN-STREAM/RIVER HABITAT CLASSIFICATION.

population

1 In ecology, a group of organisms of the same species, present in one place at one time. This definition is not always so simple to put into practice because the 'place' occupied by a population is not clear-cut. A population may extend over a large area, with limited interaction (e.g. occasional mating) between individuals in different parts of the population. POPULATION DYNAMICS nowadays stresses the metapopulation concept, where local populations form part of a larger, regional population.

2 In statistics, a term referring to all the observations that could be made on a particular item. The summary measurements of populations (PARAMETERS) are represented by Greek letters, for example, the MEAN (μ) and the variance (σ^2).
[R.H.S. & B.S.]

population bottleneck A substantial decrease in POPULATION SIZE that may be associated with a loss of genetic HETEROZYGOSITY and with SPECIATION events. [V.F.]

population cycles Changes in the numbers of individuals in a population involving repeated oscillations between periods of high and low density. [V.F.]

population density The local ABUNDANCE of a species expressed quantitatively as numbers per unit area or numbers per unit volume. Population density is more useful than POPULATION SIZE for many purposes because it provides a better indication of local abundance relative to the availability of resources.

Population size, however, provides information important in predicting FLUCTUATIONS and extinction of ENDANGERED SPECIES where populations are small in absolute terms because such species are often not widespread. Small populations are more vulnerable to random or stochastic events, INBREEDING DEPRESSION and local habitat disturbance. [R.H.S.]

population dynamics The variations in time and space of the sizes and densities of populations. The dynamics of populations are controlled by rates of birth, death, IMMIGRATION and EMIGRATION, which reflect the LIFE CYCLE of the organisms comprising the POPULATION. POPULATION SIZE and DENSITY are usually estimated by counting individuals within plots of known area or by marking individuals and recapturing them. Changes in environmental factors, existence of a seasonal breeding period, and delayed responses by limiting factors can all influence population dynamics. [V.F.]

population ecology The study of the relationships of populations of organisms with the members of the POPULATION, with other populations and with external environmental factors. *See also* ECOLOGY; POPULATION DENSITY; POPULATION DYNAMICS; POPULATION GENETICS; POPULATION GROWTH RATE; POPULATION REGULATION; POPULATION SIZE. [R.H.S.]

population genetics The branch of genetics concerned with the occurrence, patterns and changes of genetic variation in natural POPULATIONS. Population genetics combines the study of the population consequences of Mendel's laws (for instance the HARDY–WEINBERG LAW) with the study of the ecological influences that impinge upon populations and lead to NATURAL SELECTION. [G.D.J.]

population growth rate The number of individuals present at a particular time unit, divided by the number of individuals present one time unit earlier. When a POPULATION is in stable age distribution, each AGE CLASS grows in number at the same constant rate, and the growth rate equals the PER CAPITA birth rate minus the per capita death rate. [G.D.J.]

population projection matrix A matrix used to describe the dynamics of a POPULATION in which individuals are classified by age, size, developmental stage or other properties. [H.C. & C.D.]

population pyramid *See* PYRAMID OF NUMBERS.

population regulation REGULATION implies a tendency for a POPULATION to decrease if numbers are above a certain level, or conversely to increase if numbers are low. In other words, if population regulation occurs, POPULATION SIZE is controlled by density-dependent variation in birth, death, IMMIGRATION or EMIGRATION rates. Density-independent factors, such as climatic variation or random catastrophic MORTALITY, operate at a significant level in many natural populations and may obscure patterns of population regulation. Field evidence for population regulation has been hard to find in animal populations but is more common in plant populations. It is now recognized that population regulation may occur at different scales and that detection of population regulation requires study at the appropriate spatial scale. [R.H.S.]

population size The total number of individuals in a POPULATION. *Cf.* POPULATION DENSITY. [P.C.]

porewater The water that surrounds the solid particles in SEDIMENT and SOIL, also referred to as INTERSTITIAL water. [P.C.]

postzygotic conflict *See* SEXUAL CONFLICT.

potamon *See* RIVERS AND STREAMS, TYPES OF.

–3/2 power law An empirical rule, stating that the slope of a graph of the logarithm of mean plant shoot DRY MASS (on the *y*-axis) against the logarithm of plant POPULATION DENSITY (*x*-axis) is typically about –1.5. The pattern is observed in experiments on even-aged, high-density plant monocultures undergoing density-dependent MORTALITY (*see* SELF-THINNING). It is also known as Yoda's rule, or the self-thinning law. The observed constancy of slope at $-^3/_2$ is probably nothing more than a reflection of dimensional CONSTRAINTS; since population density has dimensions L^{-2} (where L = length), and plant mass has dimensions L^3, it is evident that a graph of log mass against log density should have a slope of $-^3/_2$.

The $-^3/_2$ power law is not consistent with another strong empirical relationship, the LAW OF CONSTANT FINAL YIELD, which produces a graph of log plant mass against log plant density with a slope of –1 (direct proportionality) rather than –1.5. In self-thinning, the points on the graph represent the same plant populations recorded at different times over the course of the experiment, as density-dependent mortality reduces POPULATION SIZE. In contrast, each point in graphs of the law of constant final yield represents the final BIOMASS of different populations, initiated by sowing different densities of seeds. [M.J.C.]

ppb (parts per billion) A UNIT of measurement of concentration, used especially in describing the concentrations of substances occurring in trace amounts, for example certain environmental pollutants. For example, 1 ppb = 1 part in $10^9 \equiv 1\,\mu g\,kg^{-1}$ (by mass) or $1\,\mu g\,l^{-1}$ (by volume; sometimes ppbv). Note that ppb was formerly used in the UK to mean parts per million million (i.e. 10^{12}), but US usage is now generally accepted. Where there is the possibility of doubt the unit should be defined as mass per unit mass, or mass per unit volume. [P.C.]

ppm (parts per million) A UNIT of measurement of concentration used especially in describing the concentrations of substances occurring in minute amounts. 1 ppm = 1 part in $10^6 \equiv 1\,mg\,kg^{-1}$ (by mass) or $1\,mg\,l^{-1}$ (by volume; sometimes ppmv). [P.C.]

prairie The plains GRASSLANDS of North America, named from the French word for meadow. [J.J.H.]

precocial Describing functional maturity and relative independence of the neonates at birth. Precocial (from the Latin *praecoquis* early maturing) development of young as a relative contrast to ALTRICIAL development has been recognized in mammals and birds. [J.M.S.]

predator Any heterotrophic organism that consumes any other living organism. In its narrowest sense it is often used to describe primary consumers. More broadly it may refer to PARASITES, PARASITOIDS, HERBIVORES or carnivores. Predators have a key regulatory function in natural ASSEMBLAGES and profoundly influence the structure, distribution and abundance of prey items. [R.C.]

predator–prey interactions The relationships that obtain between certain species (PREDATORS) that take advantage of other species (prey). Originally, predator–prey interactions referred exclusively to species, one of whose individuals killed those of the other for food. Soon it was recognized that such species shared many dynamic properties with others such as PARASITOIDS and their hosts, PARASITES and their hosts, and HERBIVORES and plants. In fact, the taking of individuals for food is neither necessary nor sufficient to establish the existence of a predator–prey interaction. Some food species, such as predator-resistant gastropods, actually become extinct in the absence of their consumer species, such as starfish, thus receiving a net benefit from their consumer. On the other hand, Batesian model species are damaged by Batesian mimic species although the latter never consume a single molecule of the former. [M.L.R.]

predator : prey ratio *See* TROPHIC RATIOS.

predator satiation (swamping) Reduction of the PER CAPITA loss rate to predators because of a large ratio of prey to PREDATOR individuals. Swamping occurs because each individual predator can pursue, kill and digest prey at a limited rate. [M.L.R.]

primary forest Virgin FOREST that has not been subjected to human clearance or interference. It may, however, have experienced natural catastrophes, such as fire, flood, hurricane, etc., but recovered from these and re-established a forest CANOPY. Within primary forest, there are likely to be small areas—gaps in the canopy—where minor damage has occurred or where an old tree has fallen and recovery is taking place. The primary forest should not, therefore, be regarded as a uniform cover of mature trees. [P.D.M.]

primary productivity (PP) The rate at which BIOMASS is produced per unit surface area of land or ocean by plants or chemoautotrophs. Where biomass may be described by its chemical energy, DRY MASS or carbon. 'Primary' indicates that this is the first step in the energy flow through an ECOSYSTEM; 'productivity' indicates the rate of manufacture or output. Primary PRODUCTIVITY is often, but incorrectly, used as a synonym of net primary productivity. [S.P.L.]

primary species Late-successional FOREST tree species, capable of ESTABLISHMENT in shade and/or possessing long-persistent, shade-tolerant saplings (*cf.* PIONEER SPECIES). [M.J.C.]

primary succession Primary SUCCESSION is most often defined as succession occurring on substrata that either have had no previous plant cover or that contain no organic matter. There is a problem with either definition. Succession is generally regarded as 'primary' when it occurs on a landslide where rock and subsoil, virtually free of organic matter, are exposed, despite the fact that before the landslide was formed the substratum was just beneath a developed SOIL carrying VEGETATION. Succession is also generally regarded as 'primary' where it occurs on the fresh deposits of meandering RIVERS or accreting SALT MARSHES or MANGROVE swamps, despite the fact that some remains of animals, algae and microorganisms are present in the deposits. The essential point is that primary successions start with very little organic matter in the substratum or none at all, while SECONDARY SUCCESSIONS start with some kind of developed soil. [P.J.G.]

primary woodland Woodland occupying a site which has remained wooded throughout the period since the original FORESTS of a region were fragmented. This definition refers to temperate regions. [G.F.P.]

principal component analysis A multivariate mathematical technique, which was first used as a method of ORDINATION in ecology in the early 1950s. At that time, computational difficulties prevented it from being widely used, but it is now a standard item in statistical software packages. Principal component analysis arises in many mathematical contexts and is essentially the prototype of most multivariate ordination methods. It is most easily envisaged as a means of summarizing the variation in a number of variables. [M.O.H.]

priority effect The increased competitive effect shown by one species on another species when it arrives first (has priority) at a habitat patch. [B.S.]

probability The chance of an event or measurement being observed. Probability is measured on a scale of 0 (impossible) to 1 (absolutely certain). For example, the probability, p, of getting a 'head' when you spin an unbiased coin is $p = 0.5$ (or $p = ^1/_2$, or $p = 50\%$). Probabilities can be calculated in two ways: a priori (before the event) and a posteriori (after the event). Before spinning the coin (a priori) we can imagine that the probability will be $^1/_2$ since there are two sides to the coin and we are only interested in one of them (heads). We exclude the possibility of the coin landing on its edge! After spinning the coin several times (a posteriori) we could estimate the probability by dividing the number of 'heads' actually observed by the total number of spins. The accuracy of this estimate will depend upon the total number of spins (n). For example, if by chance we get one more 'head' than we expect when $n = 10$ then $p = 60\%$, but if $n = 100$, $p = 51\%$. Some probabilities can only be estimated a posteriori because they encapsulate complex and unknown causes; for example, the probability that the egg of a particular species of bird will not hatch. [B.S.]

probability distribution A theoretical or observed distribution of probabilities or frequencies. Theoretical distributions use a priori probabilities, while observed distributions use a posteriori probabilities (*see* PROBABILITY). Some people restrict the term 'probability distribution' to the theoretical, while using the term 'frequency distribution' for the observed. However, both can be used and interpreted in the same way. The total area under a curve (for a continuous variable) or histogram (for discrete observations) is equal to one, and the area under part of the curve or histogram is equal to the probability or frequency of a particular event. [B.S.]

probability refuges Low-density REFUGES, available to an inferior interspecific competitor, that arise because of the AGGREGATED DISTRIBUTION of population numbers across habitat patches. Such refuges are an important part of the AGGREGATION MODEL OF COMPETITION. [B.S.]

product–moment correlation coefficient (*r*) A PARAMETRIC STATISTIC that summarizes the degree of CORRELATION between two sets of observations (x and y). The correlation coefficient (r) varies between +1 (perfect positive correlation), through 0 (no correlation), to −1 (perfect negative correlation). Both variables should approximately follow a NORMAL DISTRIBUTION, and are assumed to be measured with error. This is in contrast to some types of regression analysis (*see* REGRESSION ANALYSIS, CURVILINEAR; REGRESSION ANALYSIS, LINEAR), in which only the y variable (dependent variable) is considered to follow a normal distribution and to be measured with error. [B.S.]

production Sometimes a distinction is made between the process generating new BIOMASS and the output of new biomass. Some refer to the process as production and the output as PRODUCTIVITY. Occasionally the reverse senses are intended. Usually, however, the two terms are used interchangeably. [P.C.]

production efficiency (PE) The proportion of assimilated energy (A) which is incorporated into new BIOMASS (PRODUCTION, P) of an organism due to somatic GROWTH and development of reproductive tissues:

$$\frac{P}{A} \times 100$$

The rest of the assimilated energy is lost as respiratory heat. [M.H.]

productivity In ecological studies, it is common to refer to the growth and net photosynthetic rate of an ecosystem as net primary productivity (P_n or N) and gross primary productivity (P_g or G) respectively. P_g is typically presented on a time-scale of months or years and is equivalent to what plant physiologists normally call carbon dioxide (CO_2) ASSIMILATION (photosynthetic) rate. That is, it is the net rate at which CO_2 is fixed into new photoassimilate by the ecosystem. Even during the day this is substantially less than the actual rate of gross CO_2 fixation, due to photorespiratory CO_2 release, and mitochondrial CO_2 release associated with RESPIRATION by the leaves. P_n is defined as PHOTOSYNTHESIS minus plant respiration, i.e.:

$$P_n = P_g - R_p$$

where R_p is the respiration by plants, including that by leaves at night. Although it is now possible to measure P_g and R_p directly using aerodynamic and chamber gas-exchange techniques, most earlier studies concentrated on P_n and measured it as:

$$P_n = dM_p/dt + L$$

where dM_p/dt is the change in the mass of plant BIOMASS in the system studied and L represents the losses of material produced in the time studied. It has been most common to measure P_n as the change in above-ground biomass plus the loss of plant parts due to abscission and mortality (including herbivory) over a specified time interval. Thus, even though there may be no change in the net biomass of an ecosystem, P_n will be greater than zero due to litter fall and root turnover during the measurement period. [J.Ly. & J.G.]

prokaryote Lacking internal, membrane-bound organelles and a nuclear membrane. Bacteria are prokaryotes. *Cf.* EUKARYOTE. *See also* PROTOEUKARYOTE. [P.C.]

promiscuity (promiscuous breeding) A MATING SYSTEM in which members of both sexes mate or reproduce with multiple members of the opposite sex and do not form stable pair-bonds. [J.C.D.]

propagule A reproductive body (including spore, seed, bulb, cyst, EGG, bud, LARVA, etc.) that separates from the parental organism and gives rise to a new individual. The term is traditionally applied to plants and fungi, sometimes to invertebrates. Propagules often constitute the main dispersing stage in the LIFE CYCLE. [T.J.K.]

protandry A kind of SEQUENTIAL HERMAPHRODITISM in which an organism begins its reproductive career as a male, then switches sex and functions as a female. [M.T.G.]

protoeukaryote Hypothetical cell which, together with certain endosymbiotic PROKARYOTES. [P.C.]

protogyny A kind of SEQUENTIAL HERMAPHRODITISM in which an organism begins reproducing as a female, then switches sex to become a male. [M.T.G.]

province A BIOGEOGRAPHICAL UNIT, generally used in PHYTOGEOGRAPHY rather than ZOOGEOGRAPHY. [W.G.C.]

proximate–ultimate distinction How an organism responds to environmental factors can be ascribed to at least two kinds of interrelated causes. These are the immediate changes in physiology and behaviour and the control exerted on them, for example by endocrine and neural systems. There are also long-term processes that have led to the evolution of certain physiological and/or behavioural responses as ADAPTATIONS. Explanations of how physiological and behavioural events are caused by environmental factors are described as proximate (mechanistic) explanations. Explanations of how NATURAL SELECTION is likely to have favoured the evolution of those responses in the environments concerned are described as ultimate (functional) explanations. [P.C.]

prudent predator concept The idea that PREDATORS moderate their individual predation rates to avoid causing the extinction of their prey. In its extreme form, the concept requires rates that maximize the production rate of victims. Given an additive model of predation, this rate is achieved when predators maintain their victims at that population generating the point of inflection of the victim's SIGMOID GROWTH curve.

However, the evolutionary interests of individual predators usually clash with those of the predator population. If NATURAL SELECTION forces predators to become increasingly proficient as individuals, EVOLUTION will reduce populations of victims until the predators (and perhaps their victims) become extinct. In such circumstances, the evolution of prudent predators would require GROUP SELECTION. That weak force—and the more potent evolutionary opposition of their victims—protects such predators from extinction (*see* PREDATOR–PREY INTERACTIONS).

Prudent predation or an approximation of it can evolve when the FITNESS of the individual predator benefits from moderate predation. For example, the myxoma virus evolved a profoundly reduced lethality to European rabbits (*Oryctolagus cuniculus*) in Australia; lethal strains of myxoma virus kill so quickly that they have relatively few successful offspring. Predators, like giant California limpets (*Lottia gigantea*), that control, as individuals, at least a subpopulation of their victims for a considerable length of time can also evolve moderate predation rates. *Lottia* establish territories from which they exclude other grazers and within which they allow algae to flourish. Smaller, wandering limpet species rapidly denude the rocks tended by *Lottia* after it is removed experimentally. [M.L.R.]

psammon A term used to describe the communities of microscopic flora, mainly unicellular algae, and associated fauna which inhabit the INTERSTITIAL spaces between sand grains on a seashore or lakeshore. [B.W.F.]

psammosere The PRIMARY SUCCESSION associated with sand-dune formation, typically in coastal areas. It is the process whereby wind-blown sand accumulates to form dunes which become gradually larger and more stabilized by colonizing vegetation. [B.W.F.]

pseudo-arrhenotoky A haplo-diploid genetical system (*see* HAPLO-DIPLOIDY) in which both males and females arise from fertilized EGGS but where males transmit only the genome of the mother. Males become effectively haploid after the inactivation (heterochromatization) and possible elimination of the paternal genome. [M.W.S. & C.J.N.]

pseudogamy A form of PARTHENOGENESIS in which male GAMETES are required. In animals, entrance of a sperm is necessary to activate development of the egg, but the sperm pronucleus does not fuse with the egg nucleus and the male genome is never incorporated into the egg. In plants the endosperm is fertilized and contains maternal and paternal genes, but the embryo is asexual and contains maternal genes only. [M.W.S. & C.J.N.]

pseudointerference A density-dependent reduction in the per capita efficiency of a PREDATOR or PARASITOID due to a mechanism that does not involve direct behavioural interactions (INTERFERENCE). The term is frequently used to describe the reduction in efficiency with density of predators and parasitoids that concentrate on certain patches of their prey or HOST. Pseudointerference is an important stabilizing factor in many models of host–parasitoid dynamics. [H.C.J.G.]

pseudopregnancy A syndrome (also termed pseudocyesis), occurring in some mammals, in which uterine and hormonal changes following oestrus resemble certain characteristics of pregnancy but without implantation of an embryo. Pseudopregnancy frequently occurs in subordinate females of some cooperatively breeding mammals, and in these cases may have been selected as a mechanism enabling subordinate females to lactate and suckle the offspring of dominant females without producing offspring themselves, a form of helping or reproductive altruism. [J.C.D.]

pseudoreplication For statistical purposes, in making comparisons between situations it is necessary to make more than one observation per situation to consider if the variation of observations within a situation is more or less than that between situations. So observations are replicated, and it is presumed that the SAMPLING is carried out such that the replicates are independent. When they are not truly independent, pseudoreplication is said to have been effected with the likelihood of underestimating variance within situations. There are three cases common in survey work.

1 Spatial: 'upstream/downstream' comparisons around, for example, an effluent outlet pipe often involve comparisons of means from replicate samples taken either side of the POINT SOURCE. But this is not true replication because the replicates are nested within each site. Here, true replication would involve making observations on a number of similar 'upstream/downstream' sites.

2 Temporal: samples are often taken through time at the same place to test for 'before and after treatment'. But samples taken through time from the same place are likely to be correlated and therefore not independent. Again, true replication would involve comparison across a number of similar treatment situations.

3 Sacrificial: this occurs when researchers do have a number of the replicates per treatment but then inappropriately analyse units nested within experimental systems or ignore replicates by pooling data. *See also* EXPERIMENTAL DESIGN. [P.C.]

pteridophytes The ferns and their allies. [G.A.F.H.]

punctuated equilibrium A model of EVOLUTION in which a species is presumed to change rapidly as it comes into existence but quite slowly thereafter. When it subsequently undergoes SPECIATION it remains largely unchanged while producing a daughter species that diverges. Thus, the evolutionary pattern for a given species will be one of long periods of little or no change (STASIS) interrupted by periods of rapid change (punctuation) occurring during times at which speciation occurs.

Proponents of the punctuated equilibrium model of evolution argue that macroevolutionary phenomena (*see* MACROEVOLUTION) cannot be explained by the summed effects of changes of the kind that can be observed within populations (i.e. microevolutionary processes, *see* MICROEVOLUTION). In its extreme form, the punctuationist model views transitions between species as effectively instantaneous and non-adaptive, and that therefore SPECIES SELECTION is the major directing force in evolution. Others have explained the observed evolutionary patterns to arise from normalizing selection for a fixed optimum (appearing as stasis) punctuated by periods of DIRECTIONAL SELECTION—a pattern that is readily explainable by microevolutionary mechanisms.

Supporters of the punctuated equilibrium model of evolution emphasize the importance of developmental CONSTRAINTS that limit the evolution of taxa, whereas those favouring the alternative model (which is sometimes referred to as PHYLETIC GRADUALISM) emphasize the importance of selective forces, such as COEVOLUTION between competitors, between predators and prey, or between hosts and parasites. Much of the evidence for the punctuated equilibrium model derives from examination of morphological features of fossil taxa, and it is important to note that substantial morphological changes can sometimes arise from fairly simple genetic changes in critical control mechanisms of development. [V.F.]

pycnocline A zone in the water column within which the rate of change of density of seawater with increasing depth reaches a maximum. The pycnocline acts as a barrier to vertical water circulation and influences the distribution of certain chemicals in the SEA. It is often associated with oxygen minimum layers. [V.F.]

pyramid of biomass *See* ELTONIAN PYRAMID.

pyramid of energy *See* ELTONIAN PYRAMID.

pyramid of numbers If each TROPHIC LEVEL in the COMMUNITY is represented as a block whose length corresponded to the numbers in the trophic level, and if all the blocks were stacked on top of each other with the primary producers at the bottom, the result would be a characteristic pyramid-shaped structure.

The form of the pyramid varies widely between communities depending on the size of individuals. While many primary producers, such as PHYTOPLANKTON or grasses, are small relative to the size of herbivores that feed on them, so that it takes a larger number of the plants to support one animal, in other communities this is not the case: for instance, one sycamore tree provides food for hundreds of thousands of aphids. For this reason the pyramids of BIOMASS or energy are more widely used to compare the structures of different communities. [M.H.]

Q

quadrat A small area, delimited by a frame, used to study the DISTRIBUTION and ABUNDANCE of species, commonly of VEGETATION but also fauna. [A.J.W.]

quantitative genetics Continuous variation for quantitative TRAITS is caused partly by segregation of ALLELES at multiple loci (quantitative trait loci, or QTLs) affecting the trait, and partly from variation in the environment. Because the effects of the alleles at one QTL are not large relative to segregating variation at other loci and the environmental variation, classical Mendelian genetic methods for analysing genetic variation at single loci by observing segregation ratios in pedigrees cannot be used to study the genetic basis of quantitative traits. The only property of a QUANTITATIVE TRAIT that is directly observable on an individual is its measured value, or phenotypic value (P).

At the level of the POPULATION we can observe the MEAN and variance of individual phenotypic values. Populations are composed of families, so in addition we can observe the COVARIANCE of phenotypic values between different sorts of relatives. If multiple quantitative traits are measured, the CORRELATION between phenotypic values of the traits can be estimated. The goal of quantitative genetics is to determine the relationship between the observable statistical properties of quantitative traits to the unobserved properties of the genes concerned (e.g. DOMINANCE, EPISTASIS and PLEIOTROPY), and non-genetic influences. The principles of quantitative genetics are often applied in practice to plant and animal breeding and to the study of EVOLUTION, because most traits of agricultural importance (e.g. milk production in dairy cattle; yield of crops), and traits that are acted on by NATURAL SELECTION (e.g. fertility) are quantitative. [T.F.C.M.]

quantitative trait (quantitative character, metric trait) A TRAIT (character) for which variation between individuals in a population is continuous, and often normally distributed. Most morphological, physiological and behavioural characters that distinguish one individual from another are of this sort; for example, height, blood pressure, longevity. [T.F.C.M.]

quantum speciation The idea that species formation takes place by a single jump. [L.M.C.]

quartile Any of the three points that divide a DISTRIBUTION or set of observations into quarters, or four equal parts. The first quartile (Q1, or lower quartile) has 25% of observations below it. The second quartile is the median (Q2, or middle quartile) and the third quartile (Q3, or upper quartile) has 75% of observations below it. Some rounding is necessary when the number of observations is not exactly divisible by 4. *See also* PERCENTILE. [B.S.]

q_x In a LIFE TABLE the age-specific or stage-specific MORTALITY rate; the fraction of the proportion of individuals that survive to age x that die in age-class x. It is equivalent to the average probability of an individual dying at that age, and is calculated as d_x/l_x. High values of q_x indicate the age or stage with the most intense mortalities but, unlike d_x and k_x, q_x is not additive over stages. [A.J.D.]

R

r- and K-selection The terms r- and K-selection are derived from the LOGISTIC EQUATION:

$$N(t) = \frac{K}{1 + e^{c - rt}}$$

where $N(t)$ is POPULATION SIZE at time t, r is the rate of increase, c is a constant, and K is the CARRYING CAPACITY of the ENVIRONMENT. Selection that maximizes the rate of increase is termed r-selection. It has been hypothesized that there is a TRADE-OFF between selection for traits that maximize r and those that maximize K. However, evidence does not support such a hypothesis. *See also* CLASSIFICATION; EVOLUTIONARY OPTIMIZATION; LIFE-HISTORY EVOLUTION; OPTIMALITY MODELS; REPRODUCTIVE EFFORT; SELECTION PRESSURE. [D.A.R.]

radiation pollution Ionizing radiation consists of 'fast-flying' particles or waves that come naturally from the nuclei of unstable atoms or from nuclear facilities. Human beings and ecological systems are exposed daily to a certain amount of low-level radiation, or background radiation. Humans may also be exposed to radiation from medical uses. There may also be exposure due to accidental releases from the nuclear industry.

The major effects on biological systems are impacts on and consequent denaturing of macromolecules, leading in genetic material to MUTATIONS that may cause cancer. These are of considerable importance for human health, but their ecological importance is more problematic. [P.C.]

rain forest Any FOREST in either tropical or temperate latitudes that receives particularly high levels of rainfall and whose structure and composition is determined by the high humidity of the environment. The TROPICAL RAIN FORESTS lie in the equatorial regions of the world and are provided with high precipitation as a result of the development of convective air masses rising and releasing their load of water as they cool with height. The TEMPERATE RAIN FORESTS are mainly located in oceanic temperate regions, such as the Pacific north-west of North America, and parts of New Zealand and Japan, where local conditions lead to exceptionally high rainfall through much of the year. Both types of rain forest are characterized by high BIOMASS and PRODUCTIVITY. [P.D.M.]

raised bog A domed MIRE found in temperate regions. [P.D.M.]

ramet A potentially independent part of a modular plant—for example, a rooted shoot arising from a bud on a rhizome (below ground) or stolon (above ground)—that is capable of persisting when (and if) the connection to the parent plant (the rhizome or stolon) is severed. Parts of a clonal plant (or GENET), which are genetically identical except for SOMATIC MUTATION among them. [M.J.C.]

random distribution A term used to describe the spatial DISTRIBUTION of the individuals in a POPULATION. When individuals in a population follow a random distribution, the variance (s^2) of individuals per SAMPLE is equal to the MEAN number ($\bar{x}$) of individuals per sample ($s^2 = \bar{x}$). This is frequently summarized by saying that the variance to mean ratio is equal to 1 ($s^2/\bar{x} = 1$). *See also* AGGREGATED DISTRIBUTION; DISPERSION; POISSON DISTRIBUTION; REGULAR DISTRIBUTION. [B.S.]

random sample This is an unbiased subset of a POPULATION under investigation and is required for many statistical tests. A truly unbiased absolute estimate of POPULATION SIZE is only possible if the SAMPLE units are representative of the total population. The easiest way of achieving this is for each sample unit to contain a true random sample of the population. However, selecting a true random sample is not an easy task. [C.D.]

random walk model A MODEL in which variables change by finite, possibly random amounts, in random directions as time progresses. Random walk models occur in two main areas of ECOLOGY: as the underlying model for diffusive movement and as stochastic POPULATION models. For example: insect foraging movement might be modelled as a series of short steps of random length, each one in a random direction (*see also* OPTIMAL FORAGING THEORY). The change in spatial DENSITY of a population of thousands of similar insects moving in the same way could then be described by a diffusion equation model. Another example of the use of a random walk model might be as the NULL MODEL against which to test hypotheses about DENSITY DEPENDENCE: changes in POPULATION SIZE being modelled as a series of random steps up or down. [S.N.W.]

range

1 Spatial extent of species and/or POPULATION.

2 A measure of the spread or variation of a set of

observations that is simply the difference between the smallest and the largest single observation. [B.S.]

range management The utilization of extensive areas of GRASSLANDS for pastoral purposes in both temperate and tropical areas demands careful management to avoid possible destructive effects (*see* OVERGRAZING). Range management has traditionally been applied to the control of stock density of domestic animals, usually cattle or sheep, to ensure the continued productivity of the ecosystem. [P.D.M.]

rank Can refer to species in a COMMUNITY (*see* RANK–ABUNDANCE MODELS) or individuals in a social grouping (*see* DOMINANCE, SOCIAL). [P.C.]

rank–abundance models Ecological communities comprise species that vary markedly in their ABUNDANCES. Some species are extremely common, some have moderate abundance and some are rare; this is their RANK. The DISTRIBUTION of the abundances for these species can be used to characterize the DIVERSITY of a COMMUNITY more effectively than is possible by composite diversity measures such as the Shannon or Simpson indices (*see* DIVERSITY INDICES).

Four models are traditionally used to describe patterns of species abundance. These are the GEOMETRIC SERIES, the log series, the log normal and the broken stick; they represent a CONTINUUM from low to high EQUITABILITY. In certain cases, such as the log series, the models have arisen as empirical fits to widespread patterns. Other models have been devised on the basis of specific assumptions about the allocation of RESOURCES or the degree of COMPETITION amongst community members. However, the biological basis of some of the models, most notably the broken stick, has now been discredited, and in any case there is no reason why a good fit by a particular MODEL vindicates the ecological assumptions that it is based upon.

Species-abundance data can be presented in a rank–abundance plot. In these graphs the abundance of each species is plotted on a logarithmic scale against the species' rank, in order, from the most abundant to the least abundant species. Species abundances may, in some instances, be expressed as percentages to provide a more direct comparison between communities with different numbers of species. [A.E.M.]

rarity, biology of Rare species have low ABUNDANCE, or small ranges, or frequently both. Rarity is highly SCALE-dependent, both temporally and spatially, and may frequently have one apparent cause at the large scale (e.g. climatic restriction) and another at a smaller scale (e.g. soil type). Naturalists, ecologists and conservationists pay particular attention to rare species because they are more likely to become extinct and because they are considered to be of higher value. [K.T. & F.B.G.]

rarity, classification A common typology of rarity classifies plant species according to GEOGRAPHIC RANGE (large or small), local POPULATION SIZE (large or small) and HABITAT specificity (wide or narrow). Seven of the eight possible combinations of these states (i.e. in which any attribute is small or narrow) are recognized as forms of rarity. Some consider that this approach requires arbitrary and unnecessary division of too many variables, and defines rarity simply in terms of ABUNDANCE and/or range size. Other workers have defined rarity in various ways, with varying objectives, and the outcome has been to classify anything from 1% to over 90% of the species in an ASSEMBLAGE as 'rare'. Gaston proposes that the least-abundant 25% of species in an assemblage should be defined as rare.

The distinction is sometimes made between species which are 'naturally' rare and those whose rarity is a consequence of human activities. The latter are increasingly frequent in the modern world. [K.T.]

reaction norm In QUANTITATIVE GENETICS, the phenotypic expression of a GENOTYPE over different environments; i.e. the systematic change of the genotypic value of a genotype under a systematic change in environment. The reaction norm specifies the extent and type of the PHENOTYPIC PLASTICITY of a genotype. Imagine many individuals of identical genotype developing in different environments. The reaction norm is the function relating a particular environmental variable to the realized genotypic value of an individual of that genotype. Given the function that is the reaction norm, the value of the environmental variable can be regarded as the function argument, and the genotypic value of the individual the function value. The mean phenotypic value of a genotype over environments can be used to determine the reaction norm.

Reaction norms are parallel if genotypes are phenotypically plastic but differ only in average expression. If reaction norms are not parallel, genotypes that are phenotypically plastic differ in sensitivity to the environment, and GENOTYPE*ENVIRONMENT INTERACTION exists. Selection on the slope as well as on the height of the reaction norm is possible. [G.D.J.]

realized niche The NICHE in the presence of competitors and predators. Note that the realized niche of a species may differ between populations, depending on the particular species composition of the COMMUNITY. [J.G.S.]

recreation ecology The branch of ecology concerned with the impact of recreational and leisure activities on natural populations and the environment. [P.D.M.]

recruitment The addition of individuals to a POPULATION. The term is often treated as synonymous with NATALITY, but properly it also includes IMMIGRA-

TION. Net recruitment is the difference between the gains and losses of individuals from a population; in a closed population this is equivalent to births (natality) minus deaths (MORTALITY). [A.J.D.]

Red (Data) Book A catalogue, published by the IUCN (International Union for the Conservation of Nature and Nature Resources; renamed World Conservation Union), listing species that are rare or in danger of becoming extinct locally or nationally. [P.C.]

Red Queen's hypothesis The Red Queen's hypothesis, introduced by L.M. Van Valen in 1973, is that the sum of the momentary absolute fitnesses of interacting species in a BIOTA is constant. This zero-sum constraint for ecological or evolutionary change is a strong constraint; it is not meaningful in some approaches and it is counterintuitive in others. Another ambiguity applies to the name itself, which is often given without the possessive. However, the Queen herself stated her hypothesis, in *Through the Looking Glass*: 'Now here, you see, it takes all the running you can do, to keep in the same place.' [L.M.V.V.]

red tide Red, brown or yellowish discoloration of COASTAL waters during ALGAL BLOOMS of dinoflagellate PHYTOPLANKTON, due to high concentrations of accessory pigments other than chlorophyll *a*. Some Dinophyceae produce potent TOXINS that can cause illness or death of marine organisms and/or humans. [J.L.]

reforestation The replacement of FOREST cover which has been removed either recently or in the past. [J.R.P.]

refuges Locations (in time or space) where one species can escape the competitive or predatory interactions of another species. [B.S.]

refugia

1 Populations of animals and plants that survive in HABITATS of limited area in regions which would not otherwise support such organisms or communities. A refugium typically constitutes the flora and fauna of an ENVIRONMENT or COMMUNITY that was once widespread but is now considerably diminished in area. For example, the BOREAL refugia found today at the higher elevations of mountains in temperate Europe are genetically isolated representatives of the early Holocene communities of lowland western Europe. Such refugia are valued in WILDLIFE conservation. Increasingly, refugia are being created not only as a result of long-term CLIMATE CHANGE but as a consequence of climatic trends over recent decades and by rapid developments in land exploitation, particularly into tropical, and hitherto remote areas.

2 Related, but somewhat different or distinctive habitat and/or NICHE space that ameliorates COMPETITION.

[G.A.F.H.]

regeneration complex The surfaces of many RAISED BOGS bear mosaics of pools and hummocks and it was once thought that these replace one another through time in a regular sequence, pool–hummock–pool–hummock, and so on, forming a 'regeneration complex' over the MIRE surface. Many recent stratigraphic studies, however, have revealed that pools, once initiated, usually persist because DECOMPOSITION is relatively fast at the PEAT–water interface, so the concept of a regeneration complex is not currently in favour. [P.D.M.]

regeneration niche An expression of the requirements for a high chance of success in the replacement of one mature individual plant by a new mature individual of the same species. The idea requires a statement of the requirements at every stage of the LIFE CYCLE (production of viable seed, dispersal of seed, germination, establishment and onward growth). It thus embodies consideration of not only different kinds of gaps that may be filled, but also variation from year to year in the amount of seed reaching a site, germinating there and surviving both predation and physical hazards ('variable recruitment'). [P.J.G.]

regression analysis, curvilinear Many relationships in ecology can be more appropriately represented by a curved line of some kind, rather than a straight line (*see* REGRESSION ANALYSIS, LINEAR). One family of curves (of which the straight line is just one example) are known as polynomials. Three are shown below.

$y = a + bx$	straight line (linear)
$y = a + b_1x + b_2x^2$	quadratic
$y = a + b_1x + b_2x^2 + b_3x^3$	cubic

Although the first equation is traditionally called linear, it should strictly speaking be simply called the equation of a straight line. This is because mathematicians refer to all the above as linear relationships, because they are made up from a linear combination of quantities involving x. Alternatively, equations such as that for exponential increase:

$$y = ae^{bx}$$

are said to be non-linear or, more accurately, non-linear in their PARAMETERS. [B.S.]

regression analysis, linear Linear regression analysis is a technique that fits a straight-line relationship (a regression line) to a set of paired observations, using the simple straight-line equation $y = a \pm bx$. The quantity a denotes the point that the regression line crosses the y-axis (the y intercept) and the quantity b is the slope (steepness) of the regression line. If b is positive then increasing x means increasing y. If b is negative then increasing x means decreasing y.

The purpose of all regression is to enable the ecologist (or other user) to predict values of y from values of x. Ecologists may also use regression to help them explore the possible causation of changes in y

by changes in *x*, or explain some of the variation in *y* by *x*. Note, however, that the fitting of a regression line to a set of paired observations, and calling one the dependent variable, does not mean that *y* is caused by *x*. From the point of view of causation there are two fundamentally different ways of obtaining a set of paired observations. First, we could manipulate or specifically choose values of an independent variable (*x*), and subsequently measure the response of a dependent variable (*y*). This is the more usual situation for regression and is more correctly termed model I regression. Second, we could simply collect pairs of covarying observations and call one of them *y* and one of them *x*. This second situation is often more correctly analysed using CORRELATION, unless one specifically wants to predict one variable from another, when you should use a model II regression. [B.S.]

regular distribution A term used to describe the spatial DISTRIBUTION of the individuals in a POPULATION. Other equivalent terms are uniform distribution, even distribution or underdispersion. When individuals in a population follow a regular distribution, the variance (s^2) of individuals per SAMPLE is less than the MEAN number ($\bar{x}$) of individuals per sample ($s^2 < \bar{x}$). This is frequently summarized by saying that the variance : mean ratio is less than 1 ($s^2/\bar{x} < 1$). [B.S.]

regularity The type of DISPERSION pattern shown by organisms, which may be random, regular or clumped (aggregated, contagious). [A.J.W.]

regulated-escapement harvesting A form of HARVESTING that aims to ensure that the number of adults escaping harvest is sufficient to provide (potentially) satisfactory RECRUITMENT of young to the population. [J.B.]

regulated-percentage harvesting This is an uncommon term, but basically it refers to the simple idea of having a constant level of MORTALITY inflicted on the harvested population. [J.B.]

regulating factors MORTALITY (or FECUNDITY-reducing) factors which act in a density-dependent manner, such that the percentage dying increases with POPULATION DENSITY (or the proportion of realized fecundity declines with increasing population density), and this DENSITY DEPENDENCE is responsible, in interaction with density-independent factors, for determining the equilibrium POPULATION SIZE. All regulating factors are density dependent, but not all density-dependent factors are regulating. [M.J.C.]

regulation Regulation means control according to defined rules or principles, and in ecology regulation usually refers to POPULATION REGULATION. [R.H.S.]

regulator An organism that can maintain some aspect of its physiology (e.g. body temperature) constant despite different and changing properties of the external ENVIRONMENT. [V.F.]

reinforcement The evolution of increased sexual isolation among incipient species due to selection against hybrids. [J.M.Sz.]

relative growth rate (RGR) A parameter measuring the instantaneous rate of change in mass. It has the dimension of time^{-1}:

$$\mathrm{RGR} = \frac{1}{W}\frac{\mathrm{d}W}{\mathrm{d}t}$$

where W is shoot dry mass and $\mathrm{d}W/\mathrm{d}t$ is the instantaneous rate of change in shoot dry mass. RGR is typically estimated from the integral by computing the difference in log shoot mass per unit time, from two estimates of mass taken at time 0 and time t:

$$\mathrm{RGR} = \frac{\ln(W_t) - \ln(W_0)}{t} = \frac{\ln\left(\frac{W_t}{W_0}\right)}{t}$$

The longer the time period used for the assessment of RGR, the lower the maximum value of RGR that can be obtained. RGR is the preferred measure for assessing and comparing rates of GROWTH (i.e. increases in mass through time) by statistical analysis (e.g. as the RESPONSE VARIABLE for regression or ANALYSIS OF VARIANCE). [M.J.C.]

remote sensing The gathering/recording of information about a population, ecosystem, area, phenomenon, etc., using any device that is not in physical contact with the subject—for example, by radio, visually by camera, or by satellite imaging. [P.C.]

renewable energy Energy from sources that are not depleted by the process, for example solar, wind and waves. [P.C.]

renewable resource A RESOURCE that is produced by natural ecosystems at rates that can potentially balance rates of removal. [P.C.]

repeatability The closeness of agreement between successive observations employing the same method, test material and laboratory. [P.C.]

repetitive DNA Small non-coding sequences of DNA in the genome of EUKARYOTES that are expressed many times and whose function is uncertain or not entirely understood. [V.F.]

reproducibility The closeness of agreement between observations from different laboratories carrying out the same method with the same test material. [P.C.]

reproductive allocation The proportion of energy and materials available to an organism that is allocated to reproduction as opposed to GROWTH or maintenance. [W.W.W.]

reproductive effort The amount of energy that an organism channels into reproduction. For some organisms this may simply be the amount of energy that is devoted to the production of GAMETES. In many other organisms there are auxiliary patterns associated with reproduction that may consume more energy than that used in gamete production. For example, many plants produce flowers to attract pollinators, and the males of many animals expend a significant amount of energy in attracting females or fighting other males. PARENTAL CARE may also be an important fraction of reproductive effort, particularly in endothermic vertebrates. [D.A.R.]

reproductive effort model Reproductive effort is a major component of an organism's life history (*see also* LIFE-HISTORY EVOLUTION). In modelling the optimal effort it is necessary to consider the age schedule of effort and the allocation of effort per episode. [D.A.R.]

reproductive isolation The process whereby there is an absence or strong reduction of gene exchange (interbreeding) between two populations (or species) due to various, genetically controlled differences, which are also called ISOLATING MECHANISMS. Two broad categories are distinguished. Pre-mating isolating mechanisms preclude heterotypic matings. Post-mating isolating mechanisms lower the success of heterospecific matings through failure of FERTILIZATION, MORTALITY of zygotes, non-viability of F_1 hybrids, or partial or complete sterility of F_1, F_2 or back-cross hybrids. In nature, reproductive isolation between closely related species is often caused by a mixture of several pre-mating and post-mating factors. [J.M.Sz.]

reproductive synchrony The tendency of some individuals to carry out some stage of their reproductive cycle at the same time as other members of the POPULATION. [W.W.W.]

reproductive value The expected lifetime reproduction of a female of a given age (also known as Fisher's reproductive value; symbol, V). When juvenile MORTALITY is high, it may be that reproductive value peaks at an intermediate age rather than at birth. Reproductive value is calculated on the basis of age-specific survivorship (l_x) and FECUNDITY (m_x) as follows:

$$V_x = \frac{e^{rx}}{l_x}\int_x^{\infty} e^{-rt} m_t l_t \, dt$$

where r is the INTRINSIC RATE OF INCREASE, l_x is survivorship to age x and m_t is per capita fecundity at age t. Reproductive value is employed in life-history theory as a measure of the importance of individuals of different ages. It is also used in evaluating the costs and benefits of various life-history TRADE-OFFS. [M.J.C.]

reservoir hosts, parasites A reservoir HOST species is defined in respect to another host species when considering parasite transmission. The parasite is endemic in the reservoir host, but occasional transmission and consequent disease occur in the second host species. Where the second host species is man (*Homo sapiens*), the infection is said to be a zoonosis. [G.F.M.]

residence time The period of time for which an entity remains within an ecological unit of interest. [P.C.]

residual reproductive value (RRV) A concept introduced by G.C. Williams in the 1960s in the context of LIFE-HISTORY EVOLUTION. Williams suggested that REPRODUCTIVE VALUE could be partitioned into a component due to current reproduction and a component due to future reproduction. Consider a population described by an age-classified POPULATION PROJECTION MATRIX, with P_i and F_i the survival probability and fertility of age-class i. The reproductive value of age-class i (v_i) is:

$$v_i = F_i\lambda^{-1} + P_iF_{i+1}\lambda^{-2} + P_iP_{i+1}F_{i+2}\lambda^{-3} + \cdots \quad (1)$$

$$= F_i\lambda^{-1} + P_i\lambda^{-1}v_{i+1} \quad (2)$$

where λ is the POPULATION GROWTH RATE. The first of these terms is current reproduction, the second is residual reproductive value. The powers of λ account for the DILUTION EFFECT of future reproduction in a population whose size is changing. [H.C.]

resilience A measure of COMMUNITY or POPULATION stability which has many definitions. The most frequently used is the speed of recovery of a community after a DISTURBANCE from a local equilibrium. This means that a highly resilient community will return very rapidly to local equilibrium. Populations with high growth rates and communities with fewer trophic levels tend to be more resilient. A contrasting definition uses the range of conditions to which a system can be disturbed and still return to local equilibrium as a measure of resilience. Definitions of resilience are often confused with those of RESISTANCE. [C.D.]

resistance

1 The ability of an organism to show insensitivity or reduced sensitivity to a chemical that normally causes adverse effects.

2 A measure of COMMUNITY or POPULATION stability that has many different definitions. One definition is that resistance is a measure of the amount of change which can be applied to a system before it is disturbed from its equilibrium. By this definition, systems with high resistance will not move from a local equilibrium when exposed to change. An extension to this definition is to use resistance as a measure of the change of state variables within a system in response to a change in a variable. Resistance and RESILIENCE are often confused.

[P.C. & C.D.]

resource competition theory A theoretical model for POPULATION DYNAMICS and COMPETITION in which competition is modelled by explicitly relating growth and multiplication of organisms in relation to the RESOURCES they consume and compete for. Whilst originating in the microbiological literature (e.g. Tilman's diatom experiments), this line of theory has been used to analyse the determinants of competitive success and community structure in higher plants (*see* TILMAN'S MODEL OF RESOURCE UTILIZATION). There are two basic predictions of resource competition theory: the R^* rule (that the competitor which, at its monoculture equilibrium, reduces the level of a single limiting resource to the lowest level, denoted by R^*, will be competitively dominant), and the RESOURCE-RATIO HYPOTHESIS. [R.P.F.]

resource defence Resource defence can arise wherever the distribution of RESOURCES in space and/or time is relatively predictable and the benefits of defending all or part of the resource outweigh the costs of defence (the principle of economic defendability; *see* TERRITORY SIZE). Where habitats vary in resource quality, resource defence can lead to a 'despotic' distribution of competitors whereby the best habitats become occupied by the strongest competitors while weak individuals are consigned to the poorer habitats (*cf.* IDEAL FREE DISTRIBUTION). Resource defence is thus a form of INTERFERENCE COMPETITION which reduces the efficiency with which competitors exploit resources. [C.J.B.]

resource-defence polygyny A MATING SYSTEM in which males defend territories containing RESOURCES such as food or nesting sites and thereby gain mating access to females attracted to these resources. Resource-defence POLYANDRY, in which females defend resources, is rarer. [J.C.D.]

resource partitioning

1 The differential use of resource types, by species or other categories of organisms. 'COMPETITION theory' claims that resource partitioning is the major mechanism maintaining COEXISTENCE of potential competitors, referring to the principle of COMPETITIVE EXCLUSION.

2 Also refers to the partitioning of resources from PHOTOSYNTHESIS and FEEDING within organisms between metabolism and PRODUCTION.

[J.G.S.]

resource-ratio hypothesis A model proposed by D. Tilman to explain the broad-scale patterns of PRIMARY SUCCESSION. At the beginning of a primary succession there is plenty of light but no SOIL (the availability of soil nitrogen is especially low). At the end of succession there is a CLIMAX forest community in which there is very little light at the soil surface, but a relatively high rate of nitrogen cycling. Tilman's model is based on the assumption that, for any given plant species, there is a TRADE-OFF between adaptations that improve its light-gathering ability (e.g. a taller shoot system) and those that improve nutrient-gathering ability (e.g. a more wide-ranging root system). Tilman sees the process of succession as the stepwise competitive replacement of species that are good competitors for NUTRIENTS (they possess relatively small shoot systems and large root systems) by others which are better competitors for light (they have large shoot systems and relatively small root systems). [M.J.C.]

resource-utilization curve The curve defining the (relative) extent of usage across a range of RESOURCES. Resource-utilization curves have an important place in the development of theory on COMPETITION, COEXISTENCE, LIMITING SIMILARITY and species packing. [J.G.S.]

resources Materials that are essential for GROWTH and reproduction of living organisms. [M.J.C.]

respiration Respiration is the oxidation of food by organisms to release energy in the form of adenosine triphosphate (ATP; *see also* AEROBIC RESPIRATION). Most organisms require oxygen, but some bacteria use either nitrate or sulphate as the oxidant. Respiration is also equated with heat loss in the ENERGY BUDGET studies. [D.J.R. & J.I.L.M.]

respiratory quotient (RQ) The ratio of the volume of carbon dioxide (CO_2) expired to the volume of oxygen inspired over the same time. Its value reflects the substrate being used in respiratory metabolism (i.e. 1 for carbohydrates, 0.7 for fats and 0.8 for proteins) and hence is used in calculating the heat loss per oxygen inspired (oxyjoule equivalent), often used to compute heat losses in ENERGY BUDGETS. [P.C.]

response variable The y variable in statistical analysis; the response (measurement or count) for which we aim to understand the relative importance of various explanatory variables in causing variation. Formerly known as the dependent variable. [M.J.C.]

restriction map A technique for physically locating sites at which restriction enzymes cut cloned DNA. [V.F.]

reworking, of sediment The horizontal and vertical movement of SEDIMENT particles as a result of the action of water currents and of the burrowing and FEEDING activities of sediment-dwelling organisms. Reworking of sediments can complicate stratigraphic analyses because deeper sediment layers may not correspond to older layers. [V.F.]

Reynolds number (*Re*) A dimensionless quantity named after the 19th century scientist, Sir Osborne Reynolds. Fluid mechanics uses the Reynolds number to define whether the flow of fluid is laminar or turbulent. Theoretically, the Reynolds number is defined as the ratio between inertial forces and viscous forces (inertial forces are those associated with the mean motion of the fluid; viscous forces are those associated with frictional shear stress). Mathematically, the Reynolds number is defined as:

$$Re = du/v$$

where d is a characteristic length scale, u is the velocity of the fluid and v is the kinematic viscosity of the fluid. Most atmospheric flows tend to be turbulent because the magnitude of v is of the order of

10^{-5} m^2 s^{-1} and the magnitude of u is of the order of 0.1–10 m s^{-1}. [D.D.B.]

rheophilous Describing a wetland habitat, vegetation or an individual species that demands a flow of water to supply its mineral nutrient requirements; literally 'flow-loving'. [P.D.M.]

rheotrophic Describing a wetland ecosystem that receives at least part of its nutrient input through a flow of groundwater; literally 'fed by the flow'. [P.D.M.]

rhizosphere The rhizosphere is the narrow zone (*c.* 1–2 mm) of SOIL immediately surrounding the actively growing root, in which the plant has a direct influence on the soil MICROFLORA (and to some extent also, the MICROFAUNA). [J.B.H.]

richness The number of species present at a particular place and time. No assumptions are made about the relative ABUNDANCES (or EQUITABILITY) of these species. If it is not possible to enumerate all species in a COMMUNITY then it becomes necessary to distinguish between numerical species richness, which is defined as the number of species per specified number of individuals or BIOMASS, and species DENSITY, which is the number of species per specified collection area. [A.E.M.]

Richter scale A measure of seismic activity devised by C. Richter in 1935 at the California Institute of Technology. It measures how much energy is released by ground movement from the centre of a seismic shock. It is logarithmic, so a 6.0 earthquake is 10 times more powerful than a 5.0. The highest ever recording is 8.9. [P.C.]

Ricker curve An important graphical technique for the analysis of single-species POPULATION DYNAMICS, named after the Canadian fisheries biologist W.E. Ricker, who first used it. The *y*-axis shows the POPULATION DENSITY in the next generation ($t + 1$); the *x*-axis shows the population density in the present generation (t). A straight line of slope 1 passing through the origin is the replacement line, where the population in the next generation is exactly the same as the population in the present generation (i.e. $N_{t+1} = N_t$). The equilibrium population density must lie somewhere on this line. The precise location of the equilibrium depends on the intersection of the Ricker curve and the replacement line. When the Ricker curve lies above the replacement line, the population in the next generation will be higher than the population in this generation; i.e. numbers will increase. When the Ricker curve lies below the replacement line, the population will decline. The stability of the equilibrium depends upon the slope of the Ricker curve at the point of intersection. [M.J.C.]

ring species A geographical DISTRIBUTION of populations in the form of a circle overlapping at some point, in which adjacent populations interbreed except at a point where they overlap and are reproductively isolated. Where interbreeding can take place the populations appear to form a single polytypic species consisting of a series of SUBSPECIES. Where overlap and REPRODUCTIVE ISOLATION occur, however, the overlapping populations behave as good species in that they do not interbreed. [L.M.C.]

riparian Pertaining to the bank(s) of a natural watercourse. [P.C.]

risk assessment In an environmental context, this involves predicting the extent to which the potential of a substance or process to cause HARM is realized under normal and/or abnormal and/or emergency/accident conditions. For hazardous chemicals, therefore, this involves a combination of HAZARD assessment (potential to cause harm) with likely fate of chemicals in the environment and hence exposure of ecological systems. For industrial processes, risk assessment involves a combination of assessing the potential to cause problems (as a result of environmental impact and/or likelihood of litigation) with an assessment of the likelihood of those problems being manifested, which depends upon how the system is operated and managed. Ideally, this should lead to calculation of an explicit probability of a particular magnitude of effect and frequency of occurrence. But often this is not possible in environmental risk assessment due to:

- a lack of understanding of all the complex interactions;
- imprecise definition of the targets one is trying to protect.

So, risk indicators are often used instead. [P.C.]

risk aversion A typical form of preference over PROBABILITY DISTRIBUTIONS of ecological benefits or costs. The term is used commonly in foraging theory. If a forager prefers (i) increased mean food intake per unit effort and (ii) decreased food-intake variance per unit effort, then the animal is risk-averse over probability distributions of food intake. A risk-taker prefers increased levels of both mean food intake and intake variance. [T.C.]

risk management In an environmental context, the formulation and enactment of instruments that are intended to control or reduce risks to human health and the well-being of ecological systems from processes and substances. The instruments can involve regulatory (command and control) action; economic incentives, either by direct means (e.g. applying taxes or levies) or by indirect means (by providing appropriate information on performance to customers and stakeholders); and voluntary agreements. [P.C.]

river continuum concept Rivers flow, and there are bound to be changes in ecological structure and function from source (springs) to finish (ultimately the SEA). Because of the likelihood of shading from surrounding vegetation, heterotrophic processes often dominate in the upper reaches, whereas primary PRODUCTION becomes more important in the middle reaches. Because of SEDIMENT loading causing shading, heterotrophic processes are likely to become dominant again in the lower reaches. Hence there is a shift in the major sources of energy inputs along the length of a river. Moreover, the energetics of the ecosystems in any one stretch are influenced by the energetics of upstream stretches—for example by import of production and washed-out DETRITUS—and in turn influence the energetics of downstream stretches (by export). Hence there is a longitudinal CONTINUUM of ECOSYSTEM processes and COMMUNITY structure—including functional feeding types (*see* FEEDING TYPES, CLASSIFICATION). [P.C.]

river invertebrate prediction and classification system (RIVPACS) A model for assessing riverwater quality. The model was constructed empirically from an initial survey of 'clean sites'. It is based on kick samples of benthic invertebrates obtained in a specified way over a specific time. Multivariate discriminant analysis sought CORRELATION between a minimum number of physicochemical variables and taxa present. Then for a particular site the model predicts expected taxa from observations on the same physicochemical variables, and this prediction can be used as a yardstick for the taxa that are actually collected from the site. The extent of divergence between expected and observed is supposed to give an index of river quality. This kind of approach avoids PSEUDOREPLICATION but is based on correlation analysis, and so until the mechanistic basis of the correlation is defined a divergence from the expected cannot be ascribed with certainty to any particular cause. [P.C.]

rivers Rivers are the natural channels which carry water from the land surface to the oceans. Saltwater forms 97% of all the water on the planet. Of the 3% that is FRESH WATER, more than 2% occurs as ice, and of the remaining 1% two-thirds is in groundwater. The remainder comprises lake, river and atmospheric waters, of which rivers contain the smallest proportion—they account for only 0.0001% of total planetary water.

Because all components of the hydrological cycle (*see* WATER (HYDROLOGICAL) CYCLE) on the planet are linked, rivers are necessarily highly dynamic, moving water from land to OCEAN stores. In the course of this movement, rivers have the power to erode the landscape and to carry debris. Rivers carry over 90% of the debris worn from the continents and are responsible for characteristic landforms such as valleys, deltas, flood plains, meanders and braiding. Rivers are also responsible for flooding.

The area drained by a river and its tributaries is known as the CATCHMENT (UK) or WATERSHED (US). River catchments are commonly used as the basis for integrated river and land planning. [A.M.]

rivers and streams, types of A number of classification schemes have been developed to distinguish the different sections along the course of a river, with changes in the BIOTA classified with reference to the physical changes of the river.

A river arising in a mountainous region can be divided into three sections.

1 An upper (or mountain) course with a V-shaped valley in which the water flows fast enough to carry stones.

2 A middle (or foothill) course where the valley is broader in section with stable sides and less prone to erosion, and the water velocity is still fast enough to carry sand and mud in suspension.

3 A lower (or plain) course with a broad, shallow valley and a water velocity slow enough to deposit sediments.

An alternative classification system, originally designed for European rivers, divides the river into two sections.

1 Rhithron—the region extending from the source to the point where the mean monthly water temperature rises to 20°C, the oxygen concentration is always high, the water velocity is fast and turbulent, and the bed is composed of rocks, stones or gravel with a fauna that is cold stenothermic and contains no PLANKTON.

2 Potamon—the region of the river where the mean monthly water temperature rises to over 20°C, occasional oxygen deficiencies occur, the water velocity is slow and the substratum consists of either mud or sand with a fauna that is either eurythermic or warm stenothermic and does contain plankton.

In areas outside western Europe, the mean monthly water temperature cut-off point between rhithron and potamon must be adjusted according to the geographic and climatic differences.

A stream or river system consists of a pattern of tributaries joining one another and coalescing to form the main river. Tributaries can be arranged into a hierarchy. First-order streams have no tributaries, but when two first-order streams meet they form a second-order stream. When two second-order streams meet they form a third-order stream and so on. Comparing streams in different orders, there are 3–4 times as many streams, each of less than half the length and draining a little more than one-fifth the drainage area in each successive order.

Rivers can be permanent, where the water always flows and the WATER TABLE is higher (effluent streams) than the stream level; or intermittent, where the water flows only seasonally (influent streams). Both permanent and intermittent rivers may be interrupted with alternate flow on or below the surface.

Geographic areas from which rivers reach the SEA are termed exorheic, in contrast to endorheic areas, where rivers arise but never reach the sea, and arheic areas, from which no rivers arise. Rivers in endorheic areas tend to have high SALINITY levels, particularly in the lower sections of their courses. [R.W.D.]

robust dynamics Dynamics that are relatively insensitive to parameter variation or perturbations. [S.N.W.]

robustness of models MODELS are described as robust if the conclusions drawn from them are not very sensitive to the assumptions and PARAMETERS that they employ. This is important in ECOLOGY because of the difficulty of obtaining precise estimates of parameters and the necessary simplification required to make models of real systems tractable. Robustness is an attractive property in that it makes prediction possible using very uncertain information, but at the same time it makes falsification of a model difficult. [S.N.W.]

rocky shores Usually refers to the marine environment, but occasionally can refer to LAKES when there are rocky outcrops on the shoreline. In general, marine rocky shores are the most densely populated and most biologically diverse of all intertidal shores. They are characterized by prominent vertical ZONATION of the dominant species. Horizontal zonation also occurs, with the degree of wave exposure having a major influence on the types of species present.

Vertical zonation in temperate and BOREAL rocky shores generally shows a pattern with a black lichen zone at the highest level, followed by a periwinkle zone (these together are sometimes referred to as the supralittoral fringe), a barnacle zone (sometimes called the mid-littoral zone) and a lower zone dominated by various species (e.g. *Mytilus edulis*), depending upon locality (sometimes called the infralittoral fringe). The width of the various zones is controlled by the slope of the shore, the tidal range and the degree of wave exposure. Whereas the upper limit of species zones is generally considered to be controlled by physiological tolerance to ABIOTIC FACTORS (e.g. DESICCATION and temperature extremes), the lower limit is believed to be controlled mainly by interactions with predators or competitors. [V.F.]

Rosenzweig's rule An empirical rule, stating that the logarithm of net primary productivity is linearly related to the logarithm of actual evapotranspiration. [M.J.C.]

ruderal A plant characteristic of open, disturbed conditions, usually resulting from human activity. [P.D.M.]

ruminant A herbivorous animal that has a complex digestive system in which a mutualistic relationship is maintained with anaerobic CELLULOSE-digesting bacteria. [R.C.]

run-off The fraction of the precipitation falling on a land area that leaves it as surface or subsurface water flow. [D.C.L.]

S

salinity Salinity is the extent of dissolved salts in waters or soils. Salinization is the process whereby soils accumulate salts over time. Seawater contains about $33\,g\,l^{-1}$ of dissolved salts, made up as follows (all units $mg\,l^{-1}$): sodium, 10 600; magnesium, 1300; calcium, 400; potassium, 380; chlorine, 19 000; bromine, 65; sulphur, 900; carbon, 28, and many other elements at trace levels. Rainwater contains 10–50 $mg\,l^{-1}$ of dissolved salts.

Salinity of waters and soils is often assessed by measuring the electrical conductivity using a conductivity cell. SOIL is mixed with distilled water to make a paste for these measurements. The conductivity is measured in millisiemens per centimetre ($mS\,cm^{-1}$). The total dissolved solids (TDS, $mg\,l^{-1}$) in soil is closely related to the electrical conductivity (EC, $mS\,cm^{-1}$) by the following empirical relationship, provided the soil is neither very acid nor very alkaline:

$$TDS = EC \times 640$$

Soil is considered to be saline when the electrical conductivity exceeds $4\,mS\,cm^{-1}$. [J.G.]

salinization An increase in the salt concentration of the soils (or waters) of a habitat, usually resulting from human activity. The term is most frequently used of soils in dry regions which have been artificially irrigated and have become salty as a result of evaporative forces bringing the salts in solution to the SOIL surface. Sodium chloride (NaCl), calcium carbonate ($CaCO_3$), calcium sulphate ($CaSO_4$) and magnesium salts are often involved in the process. [P.D.M.]

salt flat A flat, saline area in a DESERT created by the evaporation of a former lake (playa). Often the surface is crusty and firm as a result of salt crystallization. [P.D.M.]

salt marsh A term used to describe the habitat, vegetation and fauna associated with the accretion of fine sediments (mud to fine sand) between high neap- and high spring-tide levels. Bare MUDFLATS are a typical feature below high neap-tide level, but occasionally some algae and flowering plants of the genus *Zostera* (eelgrass) may occur here. Salt marshes are globally confined to temperate to tundra latitudes, being replaced in the subtropics and tropics by MANGROVE swamps (mangals). [B.W.F.]

salt pan A shallow, isolated depression formed in a SALT MARSH and typically devoid of vascular plants. It may form from natural slight depressions of the salt-marsh surface, or from the damming of small tributary creeks. [B.W.F.]

sample A statistical term that refers to a subset or selection of all the observations that could be made on a particular item. The summary measurements of samples (statistics) are therefore estimates of population PARAMETERS. They are represented by Roman letters, for example the MEAN ($\bar{x}$) and the variance (s^2). *See also* POPULATION. [B.S.]

sampling In almost all ecological studies it will be impossible to account for every individual in a POPULATION. Therefore it is necessary to examine a subgroup of the total population and extrapolate from this to the whole population. The process by which the subgroup of the population is selected is sampling. There are several steps in the development of a sampling strategy including:

1 choice of sample unit;
2 choice of number of sample units;
3 positioning of sample units.

Random sample units are ideal in a perfectly homogeneous habitat, but in reality they are likely to produce a biased sample with an estimated variance greater than for the true population. A simple method used to minimize this problem is to take a stratified random sample: the total area is divided into equal plots and an even number of sample units are taken at random from each plot.

It might be tempting to conduct systematic sampling with sample units placed at regular intervals across a study area. There is a statistical problem with this strategy as most statistics require that a sample is taken at random from a population. However, some field ecologists suggest that estimates derived from systematic sampling are, on average, better than those from random sampling.

Timing of sampling. Most populations will be affected by season, time of day and local weather conditions. It is very important that timing is taken into account either by sampling strategy or by later analysis. [C.D.]

sampling methodology/devices Ecologists try to obtain representative samples that give indicators of the characteristics of the systems they study. One important distinction is that between relative estimates of ecological quantities (relative ABUNDANCE of species or life stages within species) and absolute estimates of ecological quantities (expressed relative to area, volume, habitat). It is therefore possible to identify broad classes of techniques that have potential applicability with respect to the two kinds of estimates. In general, those appropriate for relative estimates will not be applicable for absolute ones. One the other hand, those which are useful for absolute measures can generally be used to provide relative estimates, though the effort required would generally be prohibitive. Table S1 summarizes this classification. [P.C.]

sand dune A term used to describe a mound or hill of sand formed where wind-blown sand accretes (aeolian deposition), typically around surface obstacles, often vegetation. Sand dunes frequently form in coastal situations where specific reference can be made to embryo, fore, mobile and fixed dunes, which are components of the sand-dune SUCCESSION. [B.W.F.]

sandy shores Sandy shores can be categorized along a continuum from dissipative to reflective. Dissipative beaches experience strong wave action, have fine particles and a gentle slope, are very eroded, and most of the wave energy is dissipated in a broad surf zone. Reflective beaches tend to experience little wave action, have coarse SEDIMENT and a steep slope, are depositional, and wave energy is reflected off the beach face.

The form and community structure of sandy shores is controlled primarily by wave action, the slope of the shore and the particle size. Particle size and slope appear to be more important than wave action in controlling abundance and diversity on sandy shores. Very exposed beaches, with gentle slopes and fine grains, often are more densely populated than less-exposed beaches having coarse grains and steep slopes. Particle size strongly influences water retention, and fine sand holds more water than coarse sand. [V.F.]

saprobien system An early biotic index of water quality that recognized four stages in the oxidation of organic matter in freshwater systems: polysaprobic; α-mesosaprobic; β-mesosaprobic; and oligosaprobic. Each stage was identified by the presence/absence of INDICATOR SPECIES. The system was elaborated in the 1950s to take into account the relative ABUNDANCE of organisms in a SAMPLE. The saprobien system has been used widely in continental Europe but has not been taken very seriously in the UK or North America. [P.C.]

saprotroph (saprobe) An organism (formerly also known as saprophyte) that derives all its essential energy/NUTRIENTS from dead organic matter and therefore commonly causes the decay/biodeterioration of the latter. [J.B.H.]

satellite species Those species within a COMMUNITY which are likely to be subject to local extinctions and to periodically recolonize systems. They are likely to be both locally and regionally rare and to have weak interactions with other species in the system. These characteristics make such species likely to be the focus of conservation efforts. [C.D.]

satiation The point at which an organism no longer experiences HUNGER and stops FEEDING even when further food resources are available. [R.C.]

saturation In COMMUNITY ECOLOGY, the point reached by a local (*see* SCALE) COMMUNITY when no more species can invade from the regional species pool and establish to increase the SPECIES RICHNESS. In this situation, all RESOURCES are fully exploited and biotic interactions (particularly COMPETITION) are intense. This is not to say that other species in the regional species pool will be unable to invade and establish in a saturated community, although it is probably unlikely. If a species were to be successful it would have to be at the expense of a resident such that the total species number in the community would remain unchanged (*see* NON-INTERACTIVE COMMUNITY). Saturation, therefore, refers to the degree to which the community is able to admit new colonists from the regional species pool and the extent to which resources (niches) are exploited (*see* NON-EQUILIBRIUM COMMUNITIES). [S.J.H.]

savannah Tropical vegetation, dominated by grasses, but with various amounts of intermixed tall bushes and/or trees. [P.C.]

scale In ECOLOGY the concept of scale is largely concerned with the levels or sizes at which particular ecological entities or processes are considered. One distinction that is often made is between local, regional and biogeographic scales. Precise definitions for these distinctions are difficult to give and they often need to be interpreted

Table S1 Information that can be obtained from different sampling methods.

Sampling method	Relative abundance	Absolute abundance
Effort-based	Y	N
Time-based	Y	N
Traps	Y	S
Space-limited	Y	Y
Volume-limited	Y	Y
Habitat-limited	S	Y
Artificial substrates	S	Y
Mark–release–recapture	N	Y

Y, yes; N, no; S, some situations.

within the context of the problem being discussed.

Deciding the appropriate scales at which to view particular problems, and devising ways to conduct experiments and make observations at appropriate time- and space-scales are key issues for ecologists. It seems probable, for example, that the point on the continuum between equilibrium and non-equilibrium that a COMMUNITY occupies (*see* NON-EQUILIBRIUM COMMUNITIES) will depend on the spatial and temporal scales at which it is being viewed. Indeed, vociferous disagreements about a number of issues have arisen because conflicting results have been obtained from studies conducted at different scales. A key goal for ecologists is to find appropriate conceptual frameworks for quantifying and understanding the interactions between species and between processes which operate on different scales. It is in this latter area that hierarchical views of ecological organization seem to be making some progress. [S.J.H.]

scavenger organisms Heterotrophic organisms that specialize in FEEDING from the dead bodies of other animals but which do not undergo their whole larval development in the carcass of one animal, as do the specialist carrion-feeding invertebrates. [M.H.]

sclerophyllous vegetation SCRUB or WOODLAND vegetation in which many species have EVERGREEN, leathery leaves. Sclerophyllous literally means 'hard leaves', and the evolution of this leaf type is related to problems of water conservation. Such vegetation is, therefore, usually found in parts of the world with summer DROUGHT, such as Australia, North Africa and California. [P.D.M.]

scramble competition A term first defined by A.J. Nicholson in 1954 to describe resource use, frequently used now to describe both the behavioural process and the ecological outcome of INTRASPECIFIC COMPETITION. Nicholson defined scramble as 'the kind of COMPETITION exhibited by a crowd of boys striving to secure broadcast sweets', where 'success is commonly incomplete, so that some, and at times all, of the requisite secured by the competing animals takes no part in sustaining the population, being dissipated by individuals which obtain insufficient for survival'. Scramble competition involves passive exploitation (*see* EXPLOITATION COMPETITION) rather than direct INTERFERENCE (*see* INTERFERENCE COMPETITION). [R.H.S.]

scrub Communities dominated by shrubs or bushes which are intermediate between FOREST and GRASSLAND, often forming an intermediate zone between them. Many shrubs have, or may have, multiple stems, a tendency enhanced by coppicing. Others have a single trunk when left uncut; the distinction between shrubs and small trees is somewhat arbitrary. [J.R.P.]

sea A term often used synonymously with OCEAN or in reference to a subdivision of an ocean, or a very large enclosed body of (usually salt) water. Sea may also describe waves generated or sustained by winds within their fetch or a portion of the ocean where waves are being generated by wind. [V.F.]

sea-floor spreading The horizontal movement of oceanic crust; the process by which LITHOSPHERE is generated at divergence zones, such as midocean ridges, and adjacent lithospheric plates are moved apart as new material forms. [V.F.]

sea-level changes Alterations in the level of oceanic waters with respect to the neighbouring landmasses. Such changes are caused by the accumulation (causing levels to fall) and melting (causing levels to rise) of glaciers. Sea level is believed to have risen by about 100 m since the melting of the Pleistocene glaciers, approximately 10 000 years ago. It has been estimated that the world mean sea level has been rising at about 1.2–1.5 mm $year^{-1}$ during the first half of the 20th century. [V.F.]

searching It is assumed that foragers choose searching modes that maximize the net rate of energy gain. Two aspects have been modelled, the trajectory of the search path and the energetics of searching. [R.N.H.]

searching efficiency A parameter (symbolized, sometimes, by a) that describes the rate at which an individual PREDATOR or PARASITOID encounters prey or HOSTS. It is sometimes called the attack rate, when every encounter leads to an attack. [H.C.J.G.]

searching time In behavioural ecological models of foraging, and to a lesser degree in POPULATION DYNAMICS, a parameter (usually symbolized as T_S) that describes the time a PREDATOR or PARASITOID spends SEARCHING for prey or HOSTS. The total time available for a forager can be divided into searching time and HANDLING TIME. [H.C.J.G.]

season Any of the four equal periods, marked by the equinoxes and solstices, into which the year is divided. Because the Earth rotates and its axis is inclined in relation to the Sun, and because it revolves around the Sun in an elliptical orbit, the distribution of solar energy over the Earth's surface varies and results in changes of season. At the spring and autumn equinoxes the Equator receives maximum SOLAR RADIATION, whereas at the winter (22 December) and summer (22 June) solstices, the Sun is vertical over the Tropic of Cancer (Northern hemisphere) or the Tropic of Capricorn (Southern hemisphere). [S.R.J.W.]

secondary forest Forest (either temperate or tropical) which has regenerated following human clearance. The term may also be used in a more restricted sense implying that the land upon which the secondary forest develops has been used for alternative land use, such as pasture or arable. [P.D.M.]

secondary host For a macroparasite, a host which is parasitized but which does not actively transport the PARASITE to the definitive HOST. Asexual multiplication may occur within the secondary host, whereas the host within which sexual reproduction occurs is usually defined as the definitive or primary host. Other hosts may be vector species. [G.F.M.]

secondary productivity Productivity is the rate at which an individual population, or other ecological unit belonging to the same TROPHIC LEVEL, accumulates BIOMASS or energy by the production of new somatic and/or reproductive tissues. Productivity of AUTOTROPHS is PRIMARY PRODUCTIVITY, and productivity of HETEROTROPHS is referred to as secondary productivity, whether the heterotrophs are members of the grazing or decomposer food webs. Rates of secondary productivity are of fundamental interest because they determine the number of higher trophic levels that can be supported above them and hence the length of food chains and structure of the COMMUNITY. They are also of applied significance when evaluating the extent to which animal populations can be harvested for human use. [M.H.]

secondary sexual characteristics TRAITS which have evolved in response to differences among individuals in mating success due to COMPETITION over mates or mate choice (*see* SEXUAL SELECTION). [J.D.R.]

secondary succession This is SUCCESSION occurring where there has already been a plant cover, and there is some kind of developed SOIL. It is contrasted with PRIMARY SUCCESSION. [P.J.G.]

secondary woodland Woodland occupying a site which has at some time been artificially cleared of WOODLAND, but which has reverted to woodland by natural SUCCESSION or by planting. Clearance involves the removal of tree cover and use of the site for PASTURE, meadow, arable, building, etc., by which alternative uses a non-woodland VEGETATION has been established. [G.F.P.]

sedentary Describing an organism that remains fixed to a specific location in space throughout its lifespan. [R.C.]

sediment A general term for unconsolidated deposits of either minerogenic or organic origin. The minerogenic deposits are divisible into three groups: (i) sorted, or stratified; (ii) unsorted, or non-stratified mechanical sediments (clastic sediments); and (iii) chemical (non-clastic) sediments.

Biogenic sediments, of mainly organic origin, are divided into two main groups. AUTOCHTHONOUS sediments are formed mainly *in situ*, of material originating from the mother ecosystem. ALLOCHTHONOUS sediments consist partly of transported material.

When sediments become consolidated or lithified during the course of time, sedimentary rocks are formed, such as sandstone, shale and limestone. [Y.V.]

sedimentary cycle Any global cycle of an element in which the ATMOSPHERE plays an unimportant role, but in which the geological processes of marine sedimentation, uplift, erosion, soil weathering, and leaching dominate. Elements such as calcium and potassium have this type of cycle (whereas carbon, oxygen, nitrogen and sulphur, for instance, have important gaseous phases in their global cycling). [P.D.M.]

seed bank Seed banks are repositories of dormant seeds found either *in situ* in nature, for example SOIL SEED BANKS, or maintained artificially *ex situ* as a conservation tool or for economic uses such as crop breeding. [E.O.G. & L.R.M.]

seed dormancy A state in which the metabolism of a seed is reduced to a minimum, allowing survival under adverse conditions. [M.F.]

seed-limited recruitment Describing RECRUITMENT to a plant population when the input of extra seeds (e.g. by experimental sowing, or by a reduction in the numbers of granivorous animals) leads to an increase in the size of the reproductive plant population. In many ecosystems, adding extra seed has no impact on plant abundance because recruitment is microsite-limited (*see* DISTURBANCE) or herbivore-limited. All current theoretical models of plant POPULATION DYNAMICS are based on the assumption that recruitment is seed-limited when plant POPULATION DENSITY is low. In the field, however, herbivores may eat any additional seeds that are added experimentally, and a threshold density of extra seed might be required for SATIATION of the seed-feeding animals. [M.J.C.]

selection Usually shorthand for NATURAL SELECTION, though can also be used for ARTIFICIAL SELECTION. [P.C.]

selection arena The inevitable result of the overproduction of zygotes in COMPETITION for restricted space in the confines of a brood chamber. Selection arenas lead to disproportionate success of the fastest growing and most fit offspring, and a corresponding increase in the mean FITNESS of the brood. Hence selection arenas can be selectively advantageous to the maternal parent, especially if zygotes are very cheap to produce. The greater success of the more robust offspring in a selection arena can result from either selective abortion of slow-growing zygotes by the maternal parent, or greater provisioning of fast-growing zygotes. [C.M.L.]

selection coefficient In POPULATION GENETICS, the selection coefficient against a GENOTYPE A_iA_j is defined as:

$$s_{ij} = 1 - w_{ij}$$

where w_{ij} equals the relative FITNESS of that genotype ($0 \leq w_{ij} \leq 1$). *See also* QUANTITATIVE GENETICS. [G.D.J.]

selection gradient The selection gradient is the slope of the FITNESS PROFILE, the curve giving FITNESS as a function of a QUANTITATIVE TRAIT, either at any particular trait value or over a predetermined range of trait values. The slope of the fitness profile indicates the direction of selection on the phenotypic mean of a quantitative trait. A negative selection gradient indicates a decreasing fitness profile, and therefore selection for lower trait values. A positive selection gradient indicates an increasing fitness profile, and therefore selection for higher trait values. [G.D.J.]

selection pressure A loose term indicating that selection in a particular direction exists or that selection is of any particular strength. Selection pressure might indicate the proportion of the population that is selected to be parents of the next generation, or the SELECTION COEFFICIENT, selection differential, SELECTION GRADIENT or selection intensity. [G.D.J.]

selection response If a QUANTITATIVE TRAIT in a population is selected, the observed selection response R equals the difference between the mean trait value in the offspring population, $\bar{x}_o$, and the mean trait value in the parental population before selection, $\bar{x}_p$. Hence, $R = \bar{x}_o - \bar{x}_p$. [G.D.J.]

selective feeding The ingestion of food items in a proportion different from that available in the surrounding environment. It is usually considered as an adaptation to increase energy intake. [V.F.]

selfish DNA A gene or particular stretch of DNA that can increase in frequency in a population, despite not necessarily being advantageous, by virtue of its ability to gain over-representation (i.e. transmission at a rate greater than the Mendelian rate), rather than through causing an increase in individual FITNESS. An absence of significant levels of horizontal transmission is an implicit requirement, and the over-representation is gained typically through manipulation of the host genetic machinery (e.g. by using the machinery to make copies of itself and having these inserted elsewhere into the genome, by inhibiting gametes that do not have a copy of the same selfish DNA and a variety of other techniques).

Analogous terms include SELFISH GENE, ultra-selfish gene, selfish genetic element, self-promoting element, genetic renegade, genomic parasites and parasitic (genetic) element/DNA. These terms are sometimes used interchangeably but sometimes in contrast to each other. In its original meaning the term 'selfish gene' was intended to apply to any gene/TRAIT that spread in a population under the force of selection on that gene in favour of the trait that it coded for. It was not intended that the selfish gene must be deleterious and spread because of over-representation (this would just have been one subcategory). The usage has since been changed to that described above; i.e. one in which the ability to gain over-representation is a necessary defining feature. In the *Oxford English Dictionary* this latter meaning is given but incorrectly ascribed to Dawkins. The term 'ultra-selfish gene' has been employed to contrast with the original (Dawkins) definition of selfish gene, but is all but equivalent to the later definition (i.e. a genomic component gaining over-representation).

If a distinction exists between selfish genes (later usage, alias ultra-selfish genes) and selfish genetic elements it is that some use 'selfish gene' to include only those genes that are components of the genome (this latter term, however, tends to go undefined and has numerous grey areas). The term 'selfish genetic element' may then either be reserved for extra-genomic obligately vertically transmitted heritable elements (such as vertically transmitted bacteria found in numerous insects) or as a term to encompass both the genomic and non-genomic parasitic elements.

Selfish DNA is sometimes considered as a subclass of selfish elements, which gains over-representation through their ability to ensure that they are copied and the copies are inserted elsewhere within the genome (alias mobile DNA). This class includes a variety of classes of intron (e.g. type I introns such as the ω^+ intron of yeast mitochondrial genomes), transposable elements (a short stretch of DNA with the capability to make copies of itself), genes undergoing biased gene conversion and 'homing' elements. [L.D.H.]

selfish genes The selfish gene concept was popularized by R. Dawkins in his book *The Selfish Gene* (1976), in which he developed the argument that an organism is a gene's way of making more genes. *See also* SELFISH DNA. [V.F.]

self-limitation DENSITY DEPENDENCE in single-species models of POPULATION DYNAMICS. The key process is intraspecific density dependence. This generally takes the form of COMPETITION for limiting RESOURCES. [M.J.C.]

self-thinning A form of density-dependent plant MORTALITY that occurs during the development of an even-aged MONOCULTURE, as the individuals grow from germination to reproductive maturity. The relatively small plants die because the relatively large plants grow even bigger and shade them out. The term is sometimes used to describe any density-dependent mortality in plant populations. [M.J.C.]

semelparity Also called monocarpy (in plants) or monotely (in invertebrates), semelparity refers to a life cycle characterized by a single reproductive attempt, followed by death. According to this definition, most ANNUAL and BIENNIAL PLANTS are semelparous (some authors consider all annual and biennial species semelparous, irrespective of the length of their reproductive period). Long-lived monocarpic species include many rosette plants of deserts and the tropical subalpine zone—for example, some species of *Agave*, *Yucca* and *Lobelia*—and many bamboo (Bambuseae) species; this last group is known for its synchronous reproduction.

Semelparous species occur in most major animal taxa; in coleoid cephalopods, nereid polychaetes and mayflies (Ephemeroptera) all species are semelparous. Among vertebrates, semelparity is largely restricted to agnathans and fish, for example lampreys (*Petromyzon*), Pacific salmon (*Onchorhynchus*) and Atlantic eel (*Anguilla*). Males of the dasyurid marsupial mouse *Antechinus* are the only semelparous mammals known. [T.J.K.]

semi-natural vegetation Genuinely virgin, untouched communities are uncommon, but there is much semi-natural vegetation in which there has been minimal DISTURBANCE, thus allowing the continued existence of natural communities in a relatively unchanged form. The term is also frequently employed for communities, such as traditional hay meadows, which owe many of their characteristic properties to the way in which they have been regulated by humans, often very consistently and over long periods. [J.R.P.]

senescence

1 A synonym for AGEING. Senescence has advantages as a term for the ageing process, in that its etymology (from the Latin root *senex* or *senis*) indicates deterioration, unlike the term ageing, which can also refer to development or maturation. *See also* DISPOSABLE SOMA THEORY.

2 In botany, the deterioration of deciduous leaves at the end of the growing season, or of flowers at the end of the flowering season. The term is also used to refer to the deterioration of these structures when they undergo intermittent TURNOVER.

[M.R.R.]

sensitivity analysis

1 In general this could refer to any form of analysis that quantifies the change of one variable in response to changes in another.

2 In DEMOGRAPHY, sensitivity analysis usually refers to the sensitivity of POPULATION GROWTH RATE to changes in the vital rates.

[H.C.]

sequential hermaphroditism (sex change) The reversal of sexual modes during the life of an individual organism. Protandrous HERMAPHRODITES change from male to female mode, and protogynous hermaphrodites from female to male. Sex change is genetically programmed, but its timing may be sensitive to demographic factors, such as population SEX RATIO and survivorship. Sex change may be permanent, occurring once in the LIFE CYCLE, or reversible, occurring in each of several reproductive seasons. [R.N.H.]

sere A unit of SUCCESSION in the terminology of Clements (*see* CHARACTERS IN ECOLOGY); i.e. a kind of succession characteristic of a particular substratum and macroclimate. [P.J.G.]

serpentine soil Serpentine (serpentinite) is an ultrabasic rock dominated by the magnesium silicate minerals, antigorite and chrysotile. These rocks usually contain large amounts of nickel, which can be toxic to both plants and animals. The soils vary from deep, highly weathered soils of the tropics to shallow soils occurring mainly in the cool-temperate areas. The latter have a dark topsoil with neutral to weakly acid pH values, and the exchangeable cations are dominated by magnesium. These properties give rise to distinctive plant communities that are similar to those found on seashores. In humid tropics the soils carry a poor SCRUB forest, due largely to nickel toxicity. [E.A.F.]

sessile Describing an organism that is predominantly SEDENTARY but which is also characterized by a mobile DISPERSAL phase, as in many plants and marine organisms such as coral polyps. [R.C.]

seston Particulate matter suspended in seawater. [V.F.]

settlement A stage in the development of planktonic LARVAE during which they leave the water column to reach a suitable substratum. [V.F.]

sewage pollution POLLUTION from human excretion and defecation. Also, pollution from sewage treatment works, which may be of domestic and/or industrial origin. [P.C.]

sex allocation theory The theory that deals with the apportionment of RESOURCES by sexually reproducing parents to male and female offspring or to offspring derived from male and female GAMETES. Sex allocation includes the direct products that contribute to the next generation (i.e. gametes and offspring) as well as expenditure on accessory activities that do not contribute directly to the next generation but which may be substantial (e.g. attraction organs).

The assumption underlying sex allocation theory is that selection determines the allocation of the various structures in proportions such that the FITNESS of individuals cannot be exceeded and is evolutionarily stable. [V.F.]

sex ratios Primary sex ratio refers to the relative numbers of males and females at conception. Secondary sex ratio refers to the relative numbers of males and females at sexual maturity. Both terms are applied to organisms, usually animals, with separate sexes (i.e. non-hermaphrodites). [R.N.H.]

sex role reversal The reversal of characteristically male and female secondary sexual traits, in which females are larger, more elaborately armed or ornamented, and more aggressive than males. It occurs in a small percentage of bird, mammal, fish and insect species. Examples include many sea horses and pipe fish, spotted hyenas, and most species of jacanas. In every known case of sex role reversal, it has been found that males expend substantial time and energy caring for the young, and the phenomenon of role reversal has thus been important to the development of theories linking PARENTAL INVESTMENT and SEXUAL SELECTION. [J.C.D.]

sexual conflict Sexual conflict is said to exist if a TRAIT that increases the reproductive success of an individual of one sex can only do so at the expense of the reproductive success of an individual of the other sex. [L.D.H.]

sexual reproductive cycles Also called ALTERNATION

OF GENERATIONS, sexual reproductive cycles refer to the extent of the vegetative phase between the two steps of eukaryotic sex, fusion and segregation. These two steps of sexual reproduction are called respectively: syngamy, which doubles the amount of DNA in a cell by GAMETE fusion; and meiosis, which halves it. Three basic LIFE CYCLES are encountered among eukaryotes: diploid, haploid and haplodiploid. They differ by the relative position of syngamy and meiosis and consequently by the relative duration of the haploid and diploid phases. [M.V.]

sexual selection Charles Darwin coined the term sexual selection to explain the EVOLUTION of characters, usually possessed by one sex, that appear to be irrelevant or even disadvantageous to the survival prospects of the bearer. Darwin's view was that such characters—for instance, the long, gaudy tails of male birds of paradise or the large antlers of red deer stags—evolved to enhance the mating success of their possessors, the resulting reproductive benefit being sufficient to offset the survival cost of the character.

Sexual selection is thus sometimes regarded as a fundamentally different mechanism of evolution from classical NATURAL SELECTION. However, since natural selection favours strategies of survival only to the extent that they translate into reproductive success, the contrast is merely in the life-history component favoured rather than the mechanism of selection itself. [C.J.B.]

sexy sons hypothesis The 'sexy sons' hypothesis is a model of INTERSEXUAL SELECTION (*see* SEXUAL SELECTION), attributable to R.A. Fisher, in which female choice drives the EVOLUTION of exaggerated attractive TRAITS in males. What distinguishes it from other explanations for exaggerated traits, such as the 'handicap' hypothesis, is that 'sexy sons' traits may be entirely arbitrary and not relate to any advantageous physical quality of the male.

In Fisher's theory, females initially prefer a male trait, say a longer than average tail (it is important that the preference is relative rather than absolute), because it confers some advantage on the male, say greater agility in flight. However, the initial preference can just as easily arise because the trait is easier to detect or happens to tune in to an arbitrary sensory predisposition among females. The critical point is that the female's offspring, daughters as well as sons, inherit both the ALLELE for the preferred male trait and the allele for their mother's preference, though the trait is expressed in a sex-limited way only in sons and the preference only in daughters. [C.J.B.]

Shelford's law of tolerance An axiom, formulated by V.E. Shelford in 1913 (*see* CHARACTERS IN ECOLOGY), stating that the presence and success of an organism in a place depends upon the completeness of a complex of conditions. Alternatively, the absence or failure of an organism can be controlled by deficiency or excess with respect to any one of several factors. It reflects the fact that organisms differ in their ranges of tolerance for environmental factors. [P.C.]

shingle beach A beach made up wholly or in large measure of pebbles (SEDIMENT size 2–200 mm). [B.W.F.]

SI units (Système International d'Unités) A system of units for use in science and engineering, consisting of seven base quantities, two supplementary units, and many derived units. The base units are given in Table S2.

The supplementary units are (i) the radian (symbol: rad; the plane angle between two radii of a circle which cut off on the circumference an arc equal in length to the radius); and (ii) the steradian (symbol: sr; the solid angle which, having its vertex at the centre of a sphere, cuts off an area of the surface of the sphere equal to that of a square with sides equal to the radius of the sphere).

There are many derived units. Some of the more useful are given in Table S3.

When quantities are very small or large in relation to the base unit, they may be indicated by prefixes, as given in Table S4. For example, $1\,Mg = 10^6\,g = 1\,000\,000\,g$. Prefixes that are not multiples of 3 (h, da, d, c) are usually not necessary and should be avoided.

Current practice is that SI units are in use in all journals, but that exceptions are sometimes allowed. Chemists and soil scientists persist with eq (equivalents); Gt, or billion tonnes, is often preferred to the SI unit petagram Pg; and the litre is commonly used instead of dm^3. In forestry, the hectare (ha) is still more widely understood than multiples of m^2. Other non-SI units that are commonly allowed are minute (min), hour (h), day (d) and degree (°). *See* UNIT for other non-SI units. [J.G.]

sib-competition hypothesis Sib-competition is a mechanism that helps to explain why sexual reproduction is maintained by NATURAL SELECTION, even though the requirement to produce male offspring makes sex less efficient than ASEXUAL REPRODUCTION. The hypothesis is concerned with the ecological role of sex, i.e. with the efficient

Table S2 The seven SI base quantities, with their corresponding units and symbols.

Physical quantity	Name of unit	Symbol
Length	metre	m
Mass	kilogram	kg
Time	second	s
Electric current	ampere	A
Thermodynamic temperature	kelvin	K
Luminous intensity	candela	cd
Amount of substance	mole	mol

Table S3 Some SI derived quantities and their units.

Derived quantity	Unit name	Unit symbol	Base units
Area	square metre		m^2
Volume	cubic metre		m^3
Density	kilogram per cubic metre		$kg\,m^{-3}$
Frequency	hertz	Hz	s^{-1}
Concentration	mole per cubic metre		$mol\,m^{-3}$
Velocity	metre per second		$m\,s^{-1}$
Acceleration	metre per second per second		$m\,s^{-2}$
Force	newton	N	$m\,kg\,s^{-2}$
Pressure, stress	pascal	Pa	$N\,m^{-2}$
Viscosity (dynamic)	pascal second		Pa s
Viscosity (kinematic)	squared metre per second		$m^2\,s^{-1}$
Energy, work, heat	joule	J	N m
Power	watt	W	$J\,s^{-1}$
Electric charge	coulomb	C	A s
Potential difference	volt	V	$W\,A^{-1}$
Electrical resistance	ohm	Ω	$V\,A^{-1}$
Electrical conductance	siemens	S	Ω^{-1}
Capacitance	farad	F	$C\,V^{-1}$
Radioactivity	becquerel	Bq	s^{-1}
Irradiance, heat flux density	watt per square metre		$W\,m^{-2}$
Heat capacity, entropy	joule per kelvin		$J\,K^{-1}$
Thermal conductivity	watt per metre kelvin		$W\,m^{-1}\,K^{-1}$
Molar heat capacity	joule per mole kelvin		$J\,mol^{-1}\,K^{-1}$

Table S4 Prefixes used with SI units.

Factor	Name	Symbol
10^{18}	exa-	E
10^{15}	peta-	P
10^{12}	tera-	T
10^{9}	giga-	G
10^{6}	mega-	M
10^{3}	kilo-	k
10^{2}	hecto-	h
10	deca-	da
10^{-1}	deci-	d
10^{-2}	centi-	c
10^{-3}	milli-	m
10^{-6}	micro-	μ
10^{-9}	nano-	n
10^{-12}	pico-	p
10^{-15}	femto-	f
10^{-18}	atto-	a

exploitation of a diverse environment. It asserts that in a densely populated environment, where competition is severe, it pays to diversify.

DIVERSITY, in a heterogeneous environment, increases the number of individuals the environment can support by increasing the number of niches that can be exploited. The conditions that promote this advantage of diversity are:

1 the offspring of a single female compete with each other for survival in patches in which only a few organisms can survive and reproduce;

2 a GENOTYPE that is well adapted to one set of circumstances is less well adapted to others;

3 an environment that varies locally.

The latter two conditions are often combined as genotype by environment interactions. These conditions lead to FREQUENCY- and DENSITY-DEPENDENT SELECTION. As a consequence, a diverse range of heterogeneous organisms will outperform a mixture of CLONES, and sexual females can outcompete asexual ones. [J.C.K.]

sib selection A form of KIN SELECTION where a TRAIT is favoured because it confers an advantage on siblings. [P.C.]

sibling Any of an individual's brothers or sisters having both parents in common (full-sibling) or only one parent in common (half-sibling). [P.C.]

sibling species Morphologically similar or identical but reproductively isolated species. [P.O.]

sigmoid growth A form of limited population growth occurring where there is COMPETITION for RESOURCES. An increasing population starting at very low POPULATION DENSITY will initially show EXPONENTIAL GROWTH with an average rate of increase per individual *r*. Competition for resources will eventually cause the POPULATION GROWTH RATE to decline as population density increases. [R.H.S.]

silt A wet mixture of particles (size range 4–60 μm) found in aquatic environments and sewers, and intermediate between CLAY and mud. [P.C.]

silviculture The scientific principles and techniques of controlling, protecting and restoring the regeneration, composition and growth of natural FOREST vegetation and its plantation analogues. It is often also referred to as forestry practices. [P.M.S.A. & G.P.B.]

similarity coefficient A numerical measure, calculated from multivariate attribute data, of the similarity of two objects. Several well-known similarity coefficients increase from 0, when there is no resemblance, to 1, when there is identity.

In ecological contexts, the objects are often samples or quadrats and the attributes are species abundances. Let x_{ip}, x_{jp} be the abundances of species p in objects i, j. A popular similarity measure for such objects is Czekanowski's coefficient, defined as:

$$s_{ij} = \frac{2\sum_{p} \min(x_{ip}, x_{jp})}{\sum_{p} (x_{ip} + x_{jp})}$$

which is equal to 0 when there are no species in common, and to 1 when all species are present in equal abundance. [M.O.H.]

simultaneous hermaphroditism The simultaneous expression of male and female reproductive functions in an individual organism. [R.N.H.]

site of special scientific interest (SSSI) Any site designated by nature conservation bodies in Britain as a representative example of British habitats. Each site is seen as 'an integral part of a national series' established with the aim of 'maintaining the present diversity of wild animals and plants in Great Britain'. The selection is on scientific grounds rather than to enhance amenity or provide recreation. Powers for establishing these sites were created by the Wildlife and Countryside Act 1981. Owners and occupiers of SSSIs must seek permission to carry out any potentially damaging operation. [P.C.]

size When applied to an organism, means its dimensions. These are often expressed as length, but can be expressed in mass, volume or energy. [P.C.]

size-advantage hypothesis The hypothesis that sex change is selectively advantageous because the relationship between reproductive potential and body SIZE is different for male and female functions. Since sperm can be produced in large numbers at relatively low cost, reproductive potential as a male may at first increase rapidly with body size but then decelerate. Because EGGS are energetically relatively expensive to produce and to brood, reproductive potential as a female may remain close to zero at first, but then increase rapidly beyond a THRESHOLD body size. In such a case, the size-advantage hypothesis predicts a sex change from male to female (PROTANDRY) as the individual grows larger. [R.N.H.]

size at maturity SIZE at the start of reproductive competence. Because size at maturity is frequently related to the age at maturity and reproductive success, it is a critical TRAIT in the evolution of life histories. For females, FECUNDITY typically increases with size and hence selection acting on this aspect alone will favour an increased size at maturity. Similarly, in many species, male success is correlated with size, larger males being more successful at attracting females (or in the case of plants, pollinators) or being able to dominate smaller males. Consequently, selection will also favour larger size at maturity in males. However, the attainment of a larger size requires either an increased GROWTH rate or a longer development time. In the former case, MORTALITY may be increased because of the necessarily higher foraging rate, while in the latter case the probability of surviving to maturity and INTRINSIC RATE OF INCREASE are decreased by virtue of the increased juvenile period. Because of these conflicting effects on FITNESS there will be an OPTIMAL SIZE AT MATURITY. [D.A.R.]

size distributions A function giving the relative abundance of individuals of different sizes within a POPULATION. [H.C.]

size-efficiency hypothesis A hypothesis proposed by J.L. Brooks and S.I. Dobson to explain the shift in DOMINANCE amongst ZOOPLANKTON species in LAKES with and without fish. Generally, temperate lakes with fish are dominated by small zooplankton species, but in fishless lakes, large species are dominant. The size-efficiency hypothesis assumes that zooplankton compete for small algal cells and that the larger species are more efficient herbivores.

The hypothesis leads to three predictions.

1 Under low predation pressure (no/few fish), large zooplankton species (large cladocerans, calanoid copepods) competitively eliminate smaller species.

2 Under high predation pressure, large zooplankton are eliminated through size-selective feeding by fish; small species (rotifers, small cladocerans and copepods) dominate.

3 Under moderate predation intensity, large zooplankton populations are sufficiently reduced so as to allow COEXISTENCE of small species and enhanced DIVERSITY.

[P.S.G.]

size–number trade-off The TRADE-OFF between the SIZE and number of offspring. [R.M.S.]

size-selective predation PREDATORS generally select prey of certain size classes from amongst a range of sizes. This size-selective predation can operate within or between prey species. [P.S.G.]

skewness A statistic that measures one type of departure from a NORMAL DISTRIBUTION. It is another name for asymmetry. In skewness, one tail of the distribution is drawn out more than the other. Distributions are skewed to the right or the left depending upon whether the right or left tail is drawn out. In ecological data, distributions skewed to the right are very common. [B.S.]

social facilitation The enhancement of a behaviour by the simultaneous performance of the behaviour by a number of conspecifics or heterospecifics. [A.P.M.]

social insects Insects that form social groups, usually in the narrow sense of eusociality (*see* SOCIALITY, TYPES OF). They share the characteristics of cooperative brood care, overlap of generations (offspring stay together with parents) and the presence of a reproductive CASTE (only some members of the society reproduce). The insect order Hymenoptera contains most of the approximately 15 000–20 000 species of social insects: the ants (Formicoidea), bees (Apoidea) and wasps (Vespoidea). The order of termites (Isoptera) forms the other large group. In addition, several species belonging to various other taxa, such as sphecid wasps, aphids, and thrips, have been found to be eusocial. Although not insects, social spiders are often discussed alongside the social insects. [P.S.H.]

social organization Social organization of a POPULATION is regulated by reproductive competition among individuals and by the costs and benefits of social interactions. Such interactions lead often to TERRITORIALITY and aggression towards conspecific, competing individuals, but many species also live either permanently or temporarily in social associations. Spatial clumping of individuals can result partly from attraction of individuals to clumped resources and partly from mutual attraction of individuals to each other. Social organization can be usefully divided into AGGREGATIONS, social groups and colonies. [P.P.]

sociality, types of Sociality is the combined properties and processes of social existence. All animal species with two sexes are social, but some are more social than others: many live in societies but some do not. Social behaviour is behaviour directed differentially at members of the same species, thus including sexual behaviour. A society is usually held to be a group of cooperating individuals of the same species, with the interactions between society members involving reciprocal COMMUNICATION and going beyond sexual behaviour. [R.H.C.]

sociobiology

1 An approach to animal social behaviour which is essentially mechanistic and which regards such behaviour as subject to the same selective pressures of EVOLUTION as anatomical features and physiological processes. Social behaviour, therefore, can be explained purely in evolutionary and adaptive terms. The expression is particularly associated with the work and ideas of E.O. Wilson.

2 A logical extension is, of course, that human social behaviour can be treated in the same way. Not all would agree, especially with regard to certain aspects. The debate of NATURE VERSUS NURTURE has been at times heated here.

[P.D.M.]

soft selection *See* HARD SELECTION.

soil The term 'soil' means different things to different people. To the soil scientist, soil is the profile that is exposed when a pit is dug into the surface of the Earth (*see* SOIL PROFILE). To the agriculturist and horticulturist, soil is a medium for plant growth, and when carefully tended will produce good crops. To many others, soil is dirt; hence the term soiled. In reality, without the soil–plant system there would be no animal life on land—humans cannot exist without soil, since most food and clothing comes directly or indirectly from the land. [E.A.F.]

soil classification The CLASSIFICATION of soils is one of the most problematical areas of soil science, and despite years of effort there is still no sign of an agreed international system. There are many national systems in use at present. Here we concentrate on that of the USA.

A major innovation in *Soil Taxonomy* was the introduction of the concept of diagnostic horizons. Diagnostic horizons have quantitatively defined properties, and are used to categorize soil units. There are 26 diagnostic horizons, very briefly described as follows.

Albic: pale-coloured, leached horizon.

Argic: occurs beneath the plough layer, and has >15% coatings of HUMUS and clay.

Argillic: middle horizon with an accumulation of clay.

Anthropic: upper horizon produced by human activity.

Calcic: middle or lower horizon with an accumulation of calcium carbonate.

Cambic: middle horizon showing some degree of alteration.

Duripan: lower horizon cemented by silica.

Fragipan: hard, loamy, lower horizon with high bulk density and slakes in water.

Glossic: a horizon containing pale-coloured tongues.

Gypsic: middle or lower horizon with an accumulation of gypsum.

Histic: an accumulation of wet organic matter at the surface.

Kandic: a middle, strongly weathered, horizon with more clay than the horizon above.

Melanic: thick, black, humose upper horizon containing large amounts of allophane.

Mollic: thick, dark-coloured, surface horizon composed mainly of faunal faecal material and with high base saturation.

Natric: columnar or prismatic middle horizon with an accumulation of clay and exchangeable sodium.

Ochric: surface horizon with a low content of organic matter.

Oxic: middle and lower, highly weathered horizons with moderate to high clay content, low cation-exchange capacity and low base saturation.

Petrocalcic: hard accumulation of calcium carbonate.

Petrogypsic: hard accumulation of gypsum.

Placic: thin iron pan.

Plaggen: upper horizon formed by the continuous addition of material at the surface.

Salic: accumulation of soluble salts.

Sombric: middle horizon, containing an accumulation of translocated humus, but does not have associated aluminium, as in spodic horizons.

Spodic: middle horizon containing an accumulation of translocated humus and/or iron and/or aluminium.

Sulphuric: pH <3.5 and jarosite mottles.

Umbric: thick, dark-coloured surface horizon composed mainly of faunal faecal material, and with low base saturation.

There are also a number of diagnostic properties including: abrupt textural change; albic material; aquic conditions; coefficient of linear extensibility; durinodes; interfingering of albic materials; linear extensibility; lithic contact; *n* value; paralithic contact; permafrost; petroferric contact; plinthie; sequum and bisequum; slickensides; soft powdery lime; SOIL MOISTURE regimes; soil temperature regimes; spodic material; sulphidic material; and weatherable minerals.

The diagnostic horizons are used to ascribe soils to different orders. For example, soils with spodic horizons are Spodosols. However, it should be pointed out that some Spodosols do not have spodic horizons; likewise many Oxisols do not have oxic horizons. Thus, this system is not as precise as it appears at first; in fact there are numerous major deficiencies, hence its lack of general acceptance. [E.A.F.]

soil moisture Two aspects of SOIL moisture control its movement and availability to plants and microorganisms.

Water content. The water content of a soil is the mass or volume fraction of water that it contains.

Availability of water. The water potential is a measure of the availability of water to plant roots and microorganisms. It is the sum of the effect of the soil matrix in retaining water (the matric potential) and the effect of dissolved salts (osmotic potential). [C.E.M.]

soil profile A SOIL profile is a vertical face that is exposed when a pit is dug into the surface of the Earth. It is generally about 1 m wide, and extends down to the relatively unaltered material, which may occur at a depth that varies from <50 cm to >5 m. This vertical face usually displays a unique type of banding; the bands are known as horizons, of which there are a very large number.

Master horizons. Capital letters are used to designate master horizons. Lower-case letters are used to qualify the master horizons to subdivide the horizons.

H Accumulations of organic matter at the surface and saturated with water for long periods of the year.

O Accumulations of organic matter at the surface and not saturated with water for long periods of the year.

A An intimate mixture of organic and mineral material at or near to the surface.

E Pale-coloured medium- to coarse-textured; containing mainly quartz and resistant minerals, as a result of the leaching of CLAY, iron and aluminium.

B Has one more of the following features:

- an illuvial accumulation of clay minerals, iron, aluminium and HUMUS, alone or in combination.
- a residual concentration of sesquioxides relative to the source material.
- an alteration of the material to the extent that silicate clay is formed, oxides are liberated, or both, or granular, blocky or prismatic structures are formed.

B horizons are very variable and need to be qualified by a suffix; for example, Bt for an accumulation of clay.

C Unconsolidated material similar to the material from which the solum is formed. Strictly this is not a horizon.

R Continuous hard rock. Like the C, this is not a horizon.

[E.A.F.]

soil seed banks A soil SEED BANK is the reservoir of viable but dormant seeds which occurs in soils. [M.F.]

solar radiation The term 'solar radiation' is restricted to electromagnetic radiation emanating from the Sun, although the Sun also radiates charged particles (usually referred to as the solar wind) and neutrinos. Most of the energy of this radiation is in the wavelength range 280 nm–4 mm, but the shorter wavelength components (γ-radiation, of wavelength <0.01 nm; X-rays, of wavelength 0.01–50 nm; and ultraviolet-C radiation, of wavelength 50–280 nm—these wavelength limits are not universally agreed conventions) are sufficiently intense at the top of the Earth's ATMOSPHERE that they would kill unprotected organisms. Fortunately, all these shorter wavelength components are completely absorbed by the atmosphere. The Sun also radiates radiowaves of a wavelength exceeding 4 μm. [L.O.B.]

solarimeter An instrument for measuring the total energy radiation from the Sun, integrated over all wavelengths. [L.O.B.]

soma Early in the ONTOGENY of some multicellular organisms, cell lines giving rise to GAMETE-producing cells, tissues or organs (the germ line) become sequestered from those constituting the rest of the organism (the soma). The soma comprises the part of the organism that is affected directly by environmental influences during its life. MUTATIONS in somatic cells are not inherited by descendants. This relative independence of the germ line from direct environmental influences, dubbed Weismann's barrier, renders EVOLUTION non-Lamarckian. [D.R.B.]

somatic mutation MUTATIONS may occur either in germ line (the cells that will result in eggs and/or sperm) or in cells that will not become germ line (SOMA). Mutations in the latter are somatic mutations. Although somatic mutations may affect the viability and/or fertility of an individual, they do not affect heritable components of FITNESS as somatic alterations, by definition, will not be transmitted from parent to offspring. Numerous diseases, including various cancers, are the consequence of somatic mutations. [L.D.H.]

somatic polymorphism Although it is the case that every cell within an individual is derived from the same zygote, it need not be the case that all cells are perfectly genetically identical. The differences between the cells that are not germ line (i.e. somatic cells) constitute a POLYMORPHISM within the SOMA of an individual (cf. polymorphism within a population). [L.D.H.]

sorting, of sediment The RANGE of scatter of particle sizes about the median grain size of a SEDIMENT. Well-sorted sediment shows little variation in grain size. In general, deposits formed by wave action are better sorted than are river deposits, but not as well sorted as wind-blown sands. Glaciers form very heterogeneous unsorted deposits. The degree of sorting in a sediment influences its porosity (ratio of pore volume to total volume). [V.F.]

spawning The deposition or production of EGGS or young, usually in large numbers. [V.F.]

Spearman's rank correlation coefficient (r_s) A NON-PARAMETRIC STATISTIC that summarizes the degree of CORRELATION between two sets of observations (x and y). The correlation coefficient (r_s) varies between +1 (perfect positive correlation), through 0 (no correlation), to −1 (perfect negative correlation). The two variables need not follow a NORMAL DISTRIBUTION; it is only necessary that they can be ranked. [B.S.]

specialist A species, an individual or (in BEHAVIOURAL ECOLOGY) a STRATEGY that uses a relatively small proportion (in extreme cases only one) of the available resource types. [J.G.S.]

speciation The formation of new species. [P.C.]

species-abundance models In no COMMUNITY are all species equally abundant. Usually a few species are very common and many species are rare. Data on species ABUNDANCE can be depicted as species rank/abundance plots, showing the number of species (y-axis) falling into different abundance classes or ranks (x-axis). In addition to these graphs, species abundance data can frequently be described by one or more of a family of distributions. [B.S. & P.S.G.]

species–area relationship More species are encountered as progressively larger areas are sampled. This empirical observation is known as the species–area curve and holds whether the samples are islands of an archipelago or portions of a mainland BIOTA. The positive relationship between area and SPECIES RICHNESS is an almost ubiquitous phenomenon. [A.E.M., N.M. & S.G.C.]

species–distance relationship For an island or habitat patch of given size, equilibrium SPECIES RICHNESS is negatively related to the distance from source mainland populations. The form of the species–distance relationship depends in part on the DISPERSAL ability of species and the effectiveness of the barrier isolating island and mainland areas. [N.M. & S.G.C.]

species interactions Interactions between species can be classified and defined in various ways. A very useful method of classification is to use the 'effect' that an individual of one species has upon an individual of another species and vice versa. We ask the question: in the presence of species A (+) does species B (i) increase its numbers (+); (ii) not change its numbers (0); or (iii) decrease its numbers (−) relative to when species A is absent (−)? The same question is asked of species A in the presence of species B. The answers can be conveniently summarized in a table.

		Effect of A on B		
		+	0	−
Effect of B on A	+ 0 −	+ + 0 + − +	+ 0 0 0 − 0	+ − 0 − − −

Because of the symmetry there are only six types of interaction. These are:
0 0, neutralism;
+ 0, COMMENSALISM;
+ −, predator–prey, parasite–host, herbivore–plant interactions;
0 −, AMENSALISM;
− −, COMPETITION;
+ +, symbiosis.
See also MUTUALISM; PREDATOR–PREY INTERACTIONS. [B.S.]

species-level diversity, global Despite widespread interest in biological DIVERSITY, estimates of how many species inhabit the Earth are surprisingly vague. Larger organisms are relatively well studied and it is known, for instance, that there are about 9000 species of birds, 20 000 species of fish and 225 000 species of plants (from mosses to angiosperms). Other groups, such as marine MACROFAUNA, fungi and particularly the insects, have only begun to be catalogued. It is, however, evident that insects are by far the most speciose TAXON on Earth. At present about 1 million species of insects have been recorded. The total number of 'known' species (across all taxa) is of the order of 1.8 million. Yet, estimates based on the number of beetles in the canopies of certain tropical trees hint at much higher totals, and it has been suggested that there could be as many as 30 million tropical arthropod species. This extrapolation assumes that there are approximately 160 species of canopy beetle per tropical tree species, that beetles represent about 40% of arthropod species, that for every two insect species in the canopy there is at least one existing elsewhere on the tree, and that there are about 50 000 different species of tropical tree. Each step in the logic is fraught by uncertainty, and slight shifts in, for instance, the estimate of the proportion of the fauna specialized on a given tree species can cause the overall total to rise (to above 100 million) or fall (to below 10 million) dramatically. The most conservative estimate puts the global total of invertebrates (and hence the overall species catalogue) at 3 million.

Suggestions that other poorly studied groups, such as marine molluscs, crustaceans, polychaete worms and other benthic macrofauna may contribute another 10 million species worldwide (as opposed to the 200 000 already recorded) are open to debate. It is, however, notable that marine systems contain only about 15% of recorded species but have over 2.5 times as many phyla as terrestrial systems so may actually be much richer depositories of diversity than commonly assumed. The diversity of the microbial world is harder to quantify. Much of it is uncharted, and objective measures of microbial diversity have yet to be produced. [A.E.M.]

species richness The total number of species present in a COMMUNITY. [M.H.]

species selection A form of GROUP SELECTION in which sets of species with different characteristics increase (by SPECIATION) or decrease (by extinction) at different rates, because of a difference in their characteristics. This process is not considered INDIVIDUAL SELECTION because it involves the capacity to evolve, which is a population/species characteristic and not a property of individuals. The replacement of some species by others can cause long-term EVOLUTIONARY TRENDS and can influence the species composition of communities in which the evolved properties of species determine whether they can stably coexist. [V.F.]

specific leaf area (SLA) The area of leaf (measured on one surface only) per unit leaf dry mass. It tends to be negatively correlated with UNIT LEAF RATE (ULR). Thin leaves (high SLA) and thick leaves (low SLA) tend to occur in characteristically different light and water environments. [M.J.C.]

specific-mate recognition Specific-mate recognition is the process enabling a sexual organism to recognize another organism of opposite gender as an appropriate mate. [H.E.H.P.]

sperm competition The COMPETITION between the SPERMATOZOA from two or more males to fertilize the EGGS of a single female during one reproductive cycle. Sperm competition is a form of male–male competition that occurs between insemination and FERTILIZATION, and is a part of INTRASEXUAL SELECTION. Sperm competition has been demonstrated in a variety of insects, arachnids, crustaceans, gastropods, fish, amphibians, reptiles, birds and mammals.

Sperm competition creates opposing selective forces. On the one hand, selection favours males that succeed in fertilizing the eggs of already-mated females. On the other hand, it also favours mated males that avoid having their sperm displaced by other males. [B.B.]

sperm displacement The displacement, by sperm from a later-mating male, of sperm already present in a female's reproductive system from the site of FERTILIZATION or from the best location in the sperm storage organ(s). This is one of the principal mechanisms of SPERM COMPETITION. [P.I.W.]

spermatophore A package of sperm, usually wrapped in some protective sheath. [T.J.K.]

spermatozoa Male GAMETES, also sperm. [P.C.]

spirals of matter BIOTA remove chemical elements from the environment but return them to the environment at a later time. This leads to nutrient cycling (*see* BIOGEOCHEMICAL CYCLE). In a flowing-water system (river/stream) a nutrient released at one point will be taken up by biota downstream. So matter here is said to spiral rather than cycle. [P.C.]

sporophyte The spore-producing form of the plant LIFE CYCLE, alternating with the GAMETE-producing GAMETOPHYTE. [M.M.]

spring mire A PEAT-forming ecosystem developed over a spring head. Spring mires are RHEOTROPHIC, despite the fact that they have elevated surfaces and convex profiles, because groundwater from below is forced to their surface by pressure from the underlying spring. [P.D.M.]

stability analysis In theoretical models of POPULATION DYNAMICS, equilibrium POPULATION DENSITY (N^*) occurs under those conditions where $dN/dt = 0$ for all the interacting species. The stability analysis of a given model can be carried out graphically, analytically (by algebra) or numerically (by computer simulation). [M.J.C.]

stability, community The tendency of a COMMUNITY to return to its original state after a DISTURBANCE or to resist such disturbance. Stability may be measured in terms of the number and ABUNDANCE of its component species or by other community properties or processes. The concept of stability includes the property of RESILIENCE (the speed with which a community returns to its former state after it has been displaced from that state by a perturbation) and RESISTANCE (the ability of a community to avoid displacement in the first place). [V.F.]

stability, population According to the LOGISTIC MODEL of population growth, POPULATION SIZE is stable (i.e. is at equilibrium; POPULATION GROWTH RATE = 0) at a size corresponding to the CARRYING CAPACITY of the environment (usually symbolized by K). A population is stable if the ABUNDANCE of organisms in it is constant over time. In discrete time logistic models a population exhibits stable behaviour when the net reproductive rate, R, is such that $1 < R < 3$.

The extent to which populations fluctuate varies widely among species. Density-dependent processes (i.e. processes that cause population size to increase when below a certain level and to decrease when above a certain level) can tend to regulate population size and increase stability. Examples of density-dependent regulating mechanisms include predation, COMPETITION and PARASITISM. [V.F.]

stability–time hypothesis Deep-sea habitats have more or less constant conditions (in temperature, salinity, pressure, pH, etc.) that have existed over a long period of time. The fact that they also possess a high diversity of bivalve and polychaete species in such an apparently spatially homogeneous area, compared to the shallow, more variable, COASTAL marine environments has led to a stability–time hypothesis. This hypothesis suggests that environments that are stable over long periods of time should promote high diversity because extinctions would be rare as population FLUCTUATIONS would be low and evolutionary specialization would be encouraged, leading to a high degree of RESOURCE PARTITIONING, providing that sufficient time elapses without major change. This hypothesis has also been applied to latitudinal DIVERSITY GRADIENTS (tropics more stable and species-rich than temperate areas) and the high diversity of ancient lakes in Russia and Africa (stable over long periods of time).

However, low-diversity communities do occur in stable (but environmentally harsh) environments, and high-diversity ones in some unstable, unpredictable environments subject to intermediate degrees of DISTURBANCE (*see* INTERMEDIATE DISTURBANCE HYPOTHESIS). Also, there may be other explanations for Sanders' original data, including patchiness and HETEROGENEITY on small spatial scales in the DEEP SEA. In general, the stability–time hypothesis is unlikely to explain diversity patterns on shorter time-scales encompassing ecological interactions but is consistent with the accumulation of species diversity over evolutionary time-scales. [P.S.G.]

stabilizing selection A form of NATURAL (or ARTIFICIAL) SELECTION of a character (e.g. body mass) that selects against extreme values of the character, favouring instead some intermediate value or values. *Cf.* DIRECTIONAL SELECTION; DISRUPTIVE SELECTION. [R.M.S.]

stable age distribution The AGE STRUCTURE that is gradually approached by a population in which birth and death rates are constant. A.J. Lotka (*see* CHARACTERS IN ECOLOGY) in 1922 showed that the stable age distribution is unaffected by initial age structure. A population that has achieved stable age distribution will increase at a rate r, where r is defined by the EULER–LOTKA EQUATION. The stable age distribution can be calculated as the dominant eigenvector of the Leslie matrix (*see* POPULATION PROJECTION MATRIX). The stable age distribution is very important in population management, where the harvest derived from an exploited population (e.g. fisheries) or the damage caused by a pest population depends on the numbers in particular AGE CLASSES. *See also* DEMOGRAPHY; LIFE-HISTORY EVOLUTION; LIFE TABLE; POPULATION DYNAMICS. [R.H.S.]

stable equilibrium An equilibrium towards which a system returns after perturbation. If the perturbation is small, then return to equilibrium demonstrates the LOCAL STABILITY of the equilibrium. If the perturbation can be of any size, then the equilibrium is globally stable. [S.N.W.]

stable limit cycle A regular fluctuation in ABUNDANCE, the path of which is returned to after slight displacements from the path. It is a type of behaviour displayed by non-linear systems which involves an oscillation of stable amplitude, so that if the system starts at a point outside the limit cycle, its amplitude of oscillation declines, and if inside its amplitude increases, until the stable cycle is achieved. *See also* POPULATION DYNAMICS; POPULATION REGULATION; STABILITY, POPULATION. [V.F.]

stage distribution A DISTRIBUTION showing the relative ABUNDANCE of individuals in different developmental stages. [H.C.]

stand A general term for an area of reasonably homogeneous VEGETATION which is to be subject to scrutiny, usually for the purpose of phytosociological description. [P.D.M.]

stand cycle The name given to the cyclical dynamics exhibited by certain long-lived, patch-forming PERENNIAL plants (e.g. *Calluna vulgaris*, *Pteridium aquilinum*) where a given patch of ground passes through various stages (pioneer, building, mature and degenerate phases) in a cyclical progression. Seedling RECRUITMENT occurs only during the pioneer stage. The HABITAT as a whole comprises a mosaic of patches in different phases of the stand cycle. [M.J.C.]

standard deviation A measure of variation. In a SAMPLE it is represented by the Roman letter s, and is an estimate of the POPULATION standard deviation, represented by the Greek letter sigma (σ). It is the square root of the variance. [B.S.]

standard error (of the mean) If we take a SAMPLE of observations from a POPULATION, the sample MEAN ($\bar{x}$) will be an estimate of the population mean (μ). However, it is unlikely that $\bar{x} = \mu$ because just by chance the sample we have taken may contain a non-representative collection of observations. If we repeatedly take a large number of samples, of size n, from a population, the sample means will themselves form a DISTRIBUTION. This SAMPLING distribution of means will be normally distributed, with its mean (the grand mean of means) = μ. This result is called the CENTRAL LIMIT THEOREM. The variation in this sampling distribution of means can be measured by a STANDARD DEVIATION which, to distinguish it from the standard deviation of a sampling distribution of single observations, is called the standard error (SE) of the mean.

The SE will be influenced by two things: (i) the size of the sample (n)—smaller samples will result in more variation in the sampling distribution of means (larger SE); and (ii) the variation in the population (σ)—larger σ will give larger SE. In fact, the precise relationship, worked out by statisticians, is:

$$SE = \frac{\sigma}{\sqrt{n}}$$

In practice σ will be unknown, so we estimate the SE using the formula:

$$SE = \frac{s}{\sqrt{n}}$$

In addition to the sample mean, other sample statistics have SEs but their calculation is frequently much more complex. [B.S.]

standing crop The mass of VEGETATION in a given area at one particular time. The term is often used of above-ground vegetation only, but it can include root material also. Although most often applied to plant material, the term may also include animal BIOMASS. [P.D.M.]

startle effect Some animals when disturbed by a potential PREDATOR will expose a bright colour pattern in an attempt to startle a predator. [P.M.B.]

stasis The term evolutionary stasis is used to indicate cases in which there has been little or no apparent evolutionary change in a species for a long time. There are no agreed criteria for how long these periods of time must be or how little actual evolutionary change must occur in order for a particular species to be described as 'static'. [D.R.B.]

static life table A type of LIFE TABLE (also known as a vertical, or time-specific life table) constructed using data collected at a single time. The AGE STRUCTURE of a sample of individuals from a population is determined and is assumed to be the same as that which would have been found had a single COHORT been followed. This approach is frequently adopted for populations without DISCRETE GENERATIONS but the assumption is not true if MORTALITY rates or NATALITY rates (birth rates) are increasing or decreasing. [A.J.D.]

statistical power Statisticians frequently refer to the power of a statistical test. It is the PROBABILITY of rejecting the NULL HYPOTHESIS (H_0) when it is incorrect and the alternative hypothesis (H_1) is correct. That is, it is the probability of reaching the correct conclusion. Three things affect the power of a test.

1 The difference between the null hypothesis and the alternative hypothesis—it would be easier to distinguish between $H_0 = 0$ and $H_1 = 100$, than between $H_0 = 0$ and $H_1 = 0.01$.

2 The SAMPLE size—increasing the sample size increases the power of a test.

3 The test used—different statistical methods, testing approximately the same hypothesis, may differ greatly in their power. In general, parametric tests (*see* PARAMETRIC STATISTICS) have greater power than non-parametric tests (*see* NON-PARAMETRIC STATISTICS), when the assumptions of the parametric test are met.

See also ACCURACY/PRECISION. [B.S.]

statistical tables Tables that show the critical values of test statistics for varying levels of PROBABILITY. [B.S.]

steno- A prefix denoting narrow, particularly a narrow range of tolerance for some environmental factor; for example, stenobathic for pressure, stenothermal for TEMPERATURE, stenohaline for SALINITY. It is from the Greek *stenos* narrow. *Cf.* EURY-. [P.C.]

stenotopic Describing organisms that are only able to tolerate a narrow range of environmental conditions and hence have very restricted distributions. [M.H.]

steppe Natural GRASSLANDS of Eurasia, extending in a broad zone from Ukraine in the west to Manchuria in the east. Typical steppe is dominated by tussock-forming bunch grasses, particularly *Stipa* spp. In northern semi-humid zones, forb-rich meadow steppe forms mosaics with steppe woodlands. On the fringes of DESERT regions true steppe is replaced by semi-desert steppe containing xerophilous dwarf shrubs, particularly sagebrush (*Artemisia* spp.). In desert steppe, dwarf shrubs become dominant. [J.J.H.]

sterile male technique (sterile insect technique, SIT) A method for controlling insect pests by releasing sterilized males in large numbers. Females mate with sterilized males and the EGGS they produce are infertile. It is most effective when ratios of sterile to wild males of 10 : 1 or more can be achieved. [D.A.B.]

Stevenson screen A ventilated, wooden box used to screen thermometers from precipitation and solar and terrestrial radiation, and hence standardize the measurement of shaded air TEMPERATURE near the surface. [J.F.R.M.]

stochastic models Stochastic refers to patterns resulting from random factors. In stochastic models there is a random element such that for a given input to the model the outcome is not uniquely determined but takes a range of possible values. Thus, stochastic models predict the outcome of a process as the result of random effects and express events in terms of probabilities. [V.F.]

storage This is the retention of materials and energy by organisms rather than using these directly in GROWTH, reproduction or metabolism. It can occur outside the body, as in food caches, or within the body either throughout the tissues or in specialized depots, such as the hump of the camel. Storage may be a response to changes in the surrounding environment, for example in anticipation of food shortages associated with drought or winter conditions, and/or to changes associated with the condition of the organism, for example in anticipation of reproduction, migration or hibernation.

There are differences not only in the amount of energy that is stored but in the way that it is stored. Commonly used storage materials are polysaccharides (e.g. starches and glycogen) and lipid, although proteins are sometimes used. Lipids are more efficient energy stores than polysaccharides, in that they package more energy per gram—*c.* 40 cf. *c.* 20 J g^{-1}. However, slightly more energy is lost in the formation of lipids from fatty acids and in the transfer of energy to adenosine triphospate (ATP) than in the synthesis and utilization of polysaccharides. Glycogen is more readily available for ANAEROBIC METABOLISM. Proteins are surprisingly poor energy stores: despite having an energy density equivalent to that of polysaccharides (20–25 J g^{-1}), they incur considerably greater costs in transformation. [P.C.]

stotting A peculiar gait of African antelopes in which the legs are kept almost straight and the individual appears to bounce high off the ground from all four feet simultaneously. Stotting, as well as pronking and leaping, usually occurs in the presence of a PREDATOR, and has puzzled naturalists because it seems to be a particularly inefficient means of fleeing from attack. Recent theoretical work on honest signalling and empirical work on the antipredator behaviour of gazelles has suggested that antelopes may use stotting to signal to predators their athletic ability, and predators may choose individual prey out of a herd on the basis of stotting performance. [J.C.D.]

strandline (driftline) The line of debris left at a high-tide mark, comprising natural (mostly dead seaweed) and materials of human origin. [B.W.F.]

strategy A word often used to describe complex ADAPTATIONS—ones involving a large number of interacting and coevolving traits. It is thus applied to life-cycle and behavioural traits (*see also* EVOLUTIONARILY STABLE STRATEGY; TRAIT). Not all are comfortable with the term because it implies forethought and therefore is tainted with elements of ANTHROPOMORPHISM and TELEOLOGY. But it is now widely used and invariably neither anthropomorphism nor teleology are intended. It is sometimes contrasted with tactic—strategy being a broader, longer-term trait, but the distinction is not very sharp. For example, SEMELPARITY and ITEROPARITY might be referred to as strategies, but the reproductive investments that go with them as TACTICS. *Cf.* TACTICS. *See also* ADAPTATION; LIFE-HISTORY EVOLUTION; LIFE-HISTORY TRAITS. [P.C.]

stratification The tendency of FOREST and WOODLAND stands to develop more or less distinct layers in their vertical structure. *See also* LAKE STRATIFICATION; LAKES. [G.F.P.]

stratigraphy Stratigraphy is the study of the development and succession of stratified horizons in rocks. [P.D.M.]

stratosphere The atmospheric layer bounded below by the tropopause (10–15 km above sea level), and above by the stratopause (about 50 km above sea level), and containing 10–30% of the mass of the Earth's ATMOSPHERE. Although deformed by WEATHER systems in the underlying TROPOSPHERE, it plays little direct role in them, and contains almost no cloud except for rare mother-of-pearl clouds and diffuse polar ice clouds. Much commercial air traffic uses the low stratosphere, and the protective ozone maximum is maintained in the middle stratosphere by complex photochemistry and gentle air motion. *See also* STRATOSPHERIC CHEMISTRY. [J.F.R.M.]

stratospheric chemistry The term refers principally to the chemical processes that determine the concentration of ozone in the STRATOSPHERE. It also refers to the reactions that oxidize and decompose trace substances released into the ATMOSPHERE from natural and anthropogenic sources. If these are stable enough to escape oxidation in the TROPOSPHERE, they find their way across the tropopause into the stratosphere and are oxidized there. [P.B.]

stress The term 'stress' implies an adverse effect, but its precise definition has been somewhat elusive, largely because it is both level- and subject-dependent. It is level-dependent because an adverse response to some environmental variables at, say, a molecular or cellular level may not become manifest at an organismic/population level due to HOMEOSTASIS. It is subject-dependent because there may be GENETIC VARIANCE for stress responses between individuals in a population, and certainly between species. The application of stress to populations, as a SELECTION PRESSURE, may lead to the evolution of TOLERANCE, for example the evolution of tolerance to pesticides and pollutants. It is then debatable if a tolerant GENOTYPE or species can be described as being stressed. Even so, they are adapted to resist stress and it then becomes interesting to consider the extent to which common suites of TRAITS evolve under the same kinds of stress.

Some would restrict stress—by analogy to its definition in physics—to the environmental variables that lead to the response in biological systems—the response being referred to as 'strain'. But this is not general, and in biology the word 'stress' is used both for the cause of a response and the response itself.

Similarly, some ecologists would restrict the definition of stress to environmental factors that impair production processes—growth and reproduction—with the term 'DISTURBANCE' applied to factors that impair survival. But again this is not general. Stress is usually employed to describe impairment in general.

The concept of stress has been criticized as being unmeasurable, and as drawing an unreal distinction between favourable and unfavourable environments (the argument being that all environments are unfavourable, because the struggle for existence goes on everywhere). However, multivariate analyses of plant species' traits often produce rankings of species that reflect ADAPTATIONS to soil-nutrient availability (species appear to be ranked along a 'stress axis' from high to low nutrient availability). [P.C.]

subdominant A plant that occupies a significant proportion of the BIOMASS of an ecosystem, but is subsidiary to an even more abundant or influential dominant species. [P.D.M.]

sublittoral zone The depth zone on the shore below the level of mean low tide (or alternatively from the depth of spring low tide; *see* LITTORAL ZONE) and extending to a depth of 200 m. Also known as the subtidal zone. The vast majority of the sublittoral zone of the OCEAN consists of soft SEDIMENT. Rocky sublittoral zones, kelp forests and CORAL REEFS represent other types of sublittoral habitat, but all occur in restricted areas at relatively shallow depths. [V.F.]

submergence marsh The lower part of a SALT MARSH which is subject to (often extended) daily periods of submersion (immersion) at all TIDES, and where the key environmental factors are associated with periods of submersion (immersion). [B.W.F.]

subspecies The lowest category recognized in TAXONOMY, lying below species. Subspecies can interbreed, species cannot. Given this definition, subspecies must have distinct geographical ranges. [L.M.C.]

substitutable resource A RESOURCE which is not essential and which can be replaced by a resource of another kind. Many of the prey species consumed by a GENERALIST predator are substitutable resources. Most resources required by plants are essential, not substitutable; nitrogen cannot be substituted by phosphorus, nor phosphorus by potassium. [M.J.C.]

substrate/substratum

1 The material acted upon in a biochemical reaction.

2 The material to which organisms attach and with which they interact to a greater or lesser extent. Sometimes the substrate is used simply as a physical anchor (e.g. rocks for barnacles), whereas in other cases the substrate provides a nutrient source (e.g. SOIL for plants).

[P.C.]

subtropical The subtropical zones lie mainly polewards of 30°N and 30°S between the tropics and temperate zones. They are characterized by descending air, which is warmed as it descends and so becomes very dry. Daytime temperatures are very high as a result of intense insolation, but at night temperatures fall to near zero because clear skies allow high outgoing radiation. Rainfall is low. The subtropical zones are mainly hot deserts and the vegetation consists largely of xerophytes—PERENNIAL drought-resistant shrubs and opportunist EPHEMERALS. [S.R.J.W.]

succession Directional change in VEGETATION (or in communities of animals or microorganisms). Succession contrasts with FLUCTUATION, which occurs where change is non-directional. Formerly, some ecologists used 'succession' for any kind of vegetational change, but that usage is now rare. [P.J.G.]

successional gradient In relatively few cases the stages in a SUCCESSION occur in series along gradients of environmental conditions, for example at the ends of receding glaciers, and on large flood plains where RIVERS are meandering in such a way as to destroy the VEGETATION on one side of each bend, while depositing SEDIMENT and making it possible for succession to begin on the other side of the bend. [P.J.G.]

sulphur cycle The movement of sulphur (S) between the LITHOSPHERE (the dominant reservoir) and the ATMOSPHERE, HYDROSPHERE and BIOSPHERE, and the associated transformations between different chemical forms. [K.A.S.]

sum of squares The squared deviations of observations about a MEAN. For example, in the numerator of the formula for the variance, the 'sum of squares' = $\Sigma(\bar{x} - x)^2$. It has the important property that it can be partitioned or summed. *See also* ANALYSIS OF VARIANCE; REGRESSION ANALYSIS, CURVILINEAR; REGRESSION ANALYSIS, LINEAR. [B.S.]

sunspot cycles Sunspots are local disturbances on the solar surface, seen as dark spots when an image of the Sun is projected on to a screen. The intensity of sunspots varies on an 11-year cycle causing small variations in the energy flux to the Earth, and associated changes in the magnetosphere and ionosphere. Some people believe that these cycles influence the growth of plants and the behaviour of animals, but the effect is very small if it exists at all. *See also* MICROTINE CYCLES. [J.G.]

supergene A group of tightly linked gene loci on a chromosome which have some form of coordinated function. [P.M.B.]

superorganism A collection of single individuals that together possess the functional organization characteristic of organisms, i.e. sharing a reproduction and heredity system. Social groups of individuals of the same species can operate in this way; ECOSYSTEMS cannot. [P.C.]

superparasitism The deposition of one or a clutch of EGGS by a PARASITOID on a HOST that has already been parasitized by a member of the same species (termed self-superparasitism if the first egg or clutch is laid by the same individual). [H.C.J.G.]

supertramp species A term describing a particular pattern of DISPERSAL and colonization that may be found in certain species. Such species are characterized by high rates of dispersal across newly created habitats—which may result from processes such as terrestrial or marine island formation. Typically, such species are 'fugitives' which only persist during the early stages of colonization or on small 'islands' where more extensive communities cannot become established. [R.C.]

supply point In a system with two potentially limiting RESOURCES, the supply point is the point in phase space defined by the maximal amounts of two resources in a given habitat. The rate of supply of a resource is assumed to be proportional to the difference between the supply (S) and the availability (R) of the resource: i.e. $dR/dt = a(S - R)$. This defines a resource supply vector which always points directly to the supply point, and which is used to determine the point on the ZERO GROWTH ISOCLINE at which equilibrium occurs. In a two-resource system, this will be at the point where a species is equally limited by both resources. [M.J.C.]

supply-side ecology An investigative approach to examining factors that influence local POPULATION DYNAMICS and COMMUNITY composition and structure, which stresses the importance of the supply of colonists or RECRUITMENT (often involving physical transport processes) from a regional pool. The term has largely been developed in relation to marine intertidal communities. These communities are composed largely of SESSILE adults, the competitively dominant species of which have PELAGIC larvae for at least part of their life cycle (e.g. barnacles). In many intertidal communities, SPECIES RICHNESS and community processes appear to be largely determined by the supply of recruits and SETTLEMENT out of the water column on to the hard substrate. Recruitment is, in turn, influenced by offshore processes such as predation on larvae, larval growth rates and physical transport processes such as currents.

Supply-side ecology can also potentially be applied to PARASITE communities, where availability of parasites could play a major role in determining parasite communities in the HOST species or individual. Supply-side ecology has also been considered in the context of island communities and has a logical counterpart in terrestrial ecology, where mechanisms of PROPAGULE introduction and community fragmentation become important. *See also* DISPERSAL; ISLAND BIOGEOGRAPHY. [P.S.G.]

surplus power The energy remaining after standard and active metabolism have been subtracted from the total ENERGY BUDGET. This fraction of the energy budget can be allocated to increased growth and/or reproduction. [D.A.R.]

survival rate The proportion of the organisms at one age or stage remaining alive at the beginning of the next. It is equivalent to the average probability of an individual staying alive from one age to the next. [A.J.D.]

survivorship curve A graph of the numbers of an organism against time or age; particularly as produced by plotting a_x, l_x or $\log l_x$ from a LIFE TABLE against x. Use of log abundances produces curves with the same gradient for populations where the MORTALITY rates are the same. Curves derived from COHORT LIFE TABLES or from STATIC LIFE TABLES for populations with constant mortality and NATALITY rates will always have negative slope, and the shape of the curve will describe the distribution of mortality with age, mirroring plots of mortality (e.g. q_x or k_x) against age. [A.J.D.]

suspended solids Non-living organic particles occurring in the water column of flowing waters, LAKES or the marine environment, and maintained in suspension by physical forces. [P.C.]

suspension feeding A means of obtaining food by intercepting particles from the surrounding water. [V.F.]

sustainable agriculture Methods of agriculture that do not involve HARM to the farm as an ECOSYSTEM and/or to surrounding ecosystems. [P.C.]

sustainable development Development that meets the needs of the present generation without compromising the ability of future generations to meet their own needs. It is neither an obvious logical necessity, nor a principle that is tried and tested. Indeed, it provides much room for interpretation, which is why many pages have been written on it and almost as many clarifying definitions have been offered in place of that given above. It is in reality a statement of faith that we can live in some kind of balance with our environment. [P.C.]

sustainable yield Removing RESOURCES from the environment (HARVESTING) at a rate which allows balanced replacement by natural processes. The return per unit harvest effort will be optimized at or near this level. However, despite much effort at improving population models, ecologists are still not usually in a position to precisely define sustainable YIELD. So very often the precautionary principle is employed, for example in defining fishing quotas. [P.C.]

swamp The definition of this type of WETLAND varies in different parts of the world, but one important characteristic is that the SEDIMENT surface normally remains below the WATER TABLE throughout the year. [P.D.M.]

sweepstake dispersal route The most hazardous and unpredictable of the kinds of pathway that may be available for the migration of an organism from one habitat to another.

A sweepstake route is the least favourable of potential migration pathways in that it can only be traversed very infrequently and with great difficulty, or only under very rarely occurring circumstances. It is most frequently used in the context of a pathway leading to the colonization by organisms of an isolated island, where a strong element of chance is involved in any potential colonizer actually arriving from some remote source and establishing itself. A biota which has successfully colonized an isolated island or group of islands by means of sweepstake dispersal will typically have quite diverse elements in it. [W.G.C.]

switching The ability of a PREDATOR or PARASITOID to change from attacking one species of prey or HOST to another species as the density of the latter increases or the former declines. [H.C.J.G.]

symbiont An organism that lives in close association with another organism. The relationship is most frequently mutually beneficial (mutualistic) as each 'selfish' individual derives a direct benefit from the association. [R.C.]

symmetric competition A term that is generally synonymous with SCRAMBLE COMPETITION, but used most commonly to describe this form of competition in plant populations. Under this mode of competition, no individuals achieve competitive dominance. Instead, RESOURCES are divided more evenly amongst competing individuals. [R.P.F.]

sympatric speciation A mode of non-allopatric speciation which occurs in the absence of geographic isolation when sister species evolve within the DISPERSAL range of the offspring of a single DEME. [G.L.B.]

systems ecology The application of systems theory to ecological processes and entities; i.e. an approach based on the premise that these need to be studied and represented as integrated wholes. The study should be 'top down', because study of parts alone (i.e. reductionism) misses the all-important interactive components. Such an approach has involved the application of control theory, information theory, black-box analysis, etc. [P.C.]

T

***t* test** This is the small SAMPLE (either n_1 or $n_2 < 30$) version of the $z(d)$ TEST. Basically, the formula for the test is the same, except that we treat the result as t rather than $z(d)$. [B.S.]

tactics A set of coadapted TRAITS that solve particular ecological problems. [D.A.B.]

taiga A Russian term denoting a marshy, Siberian WOODLAND and often used to denote the circumpolar BOREAL FOREST. However, it is also used to describe the ECOTONE zone between the boreal forest and the Arctic TUNDRA. [R.M.M.C.]

taxon (pl. taxa) The organisms or species that fill a systematic category. The Linnaean taxonomic units are species, genera, families, orders, classes and phyla. Many lineages also include finer divisions such as suborder and superfamily. Taxonomic units above the species level are arbitrary, depending on the attitude of the systematist and the vagaries of EVOLUTION in a LINEAGE. [D.A.B.]

taxonomic categories Biologists recognize a hierarchy of taxonomic categories that expresses the relatedness between organisms. Groups of organisms showing a few major features in common can be subdivided into groups showing these plus other features, etc. Linnaeus thought that this represented a scale of order imprinted on nature by God and specified a series of categories arranged hierarchically to define it, viz:

Kingdom
 Class
  Order
   Genus
    Species

Moving down the series, the categories contain more and more characters in common. Through the years many additional categories have come to be used, but the system is still based on the Linnean hierarchy, although this is now viewed as an outcome of evolutionary processes. Most modern biologists recognize the following categories:

Kingdom
 Phylum (usually Division in botany)
  Class
   Order
    Family
     Genus
      Species

Each category can be subdivided further. Thus at the level of order we can have: superorder, order, suborder, infraorder. [P.C.]

taxonomy Taxonomy is the theory and practice of describing the DIVERSITY of organisms and of ordering this diversity into a system of words, called CLASSIFICATIONS, that convey information concerning kinds of relationships among organisms. In addition, taxonomy encompasses the theory and history of classification. Practitioners of taxonomy are called taxonomists and: (i) make inventories of the world's BIODIVERSITY; (ii) describe and name the species representing that biodiversity; (iii) classify those species in a system of nomenclature; (iv) revise those classifications according to new information and new methods of classification; and (v) build and maintain databases for storing and retrieving all known information about those species. [D.R.B.]

Taylor's power law This index of DISPERSION makes use of the fact that in natural populations the variance is not independent of the MEAN. In most biological populations variance increases with increasing mean. In order to calculate Taylor's power law a series of sets of samples are required. The sets of samples may be of different sizes or from different areas. What is required is a range of SAMPLE means to investigate the relationship between mean and variance. Taylor's power law states that the relationship between mean and variance can be described as follows:

$$s^2 = am^b$$

where a and b are constants, s^2 is the variance and m is the mean.

It follows that there is a linear relationship between log(mean) and log(variance):

$$\log s^2 = \log a + b \log m$$

The constant a is largely a SAMPLING factor and is usually ignored. The slope of the relationship (b) can be used as an index of dispersion. If populations follow a random (Poisson) distribution, the variance will be equal to the mean at all densities and therefore the slope of the line (b) will be 1. An aggregated or clumped population will have a slope greater than 1, while in an ordered population the slope will be less than 1.

Taylor's power law holds true for most populations and therefore provides an excellent dispersion index. One drawback is that multiple sets of samples are required which requires greater effort. Another possible drawback is that the relationship between log(mean) and log(variance) is not always exactly linear. *See also* INDICES OF DISPERSION. [C.D.]

telemetry The automatic transmission of data from remote sites. Data may be sent by landline, often the public telephone network, or transmitted by radio. [J.H.C.G.]

teleology This comes from the Aristotelian Fourth Cause and implies goal directedness, design in nature, even the future in some way influencing processes of the present. Organisms appear teleological with general 'goals' being provided by NATURAL SELECTION in terms of FITNESS. And there is a sense in which future goals are programmed into the genome of organisms by natural selection, with control systems that direct the processes and behaviour towards these ends. However, scientists do not accept conscious design in nature except when it emanates from the consciousness of humans. [P.C.]

temperate deciduous forest These are FORESTS whose trees reduce water loss during the cold winter season by dropping their leaves. Such ecosystems are often termed 'summer-green' or 'cold-deciduous' forests. They are usually strongly stratified with distinct tree, shrub, herb/dwarf shrub and cryptogamic layers, the latter being rich in mosses and liverworts. Lichens tend to be less common in the ground cover than in BOREAL FORESTS. [J.R.P.]

temperate grassland Grass- (Poaceae) or sedge- (Cyperaceae) dominated vegetation found in temperate climatic zones. In the Northern hemisphere such GRASSLANDS occur over large continuous belts of North America and Eurasia. In the Southern hemisphere they are found more discontinuously, principally in southern Africa, South America and Australasia. A defining characteristic is the abundance of cool-season C_4 grasses of the subfamilies Arundionoidae and Pooideae in temperate grasslands. These are replaced in tropical and subtropical grasslands by warm-season C_3 grasses of the subfamilies Chloridoideae and Panicuoideae. [J.J.H.]

temperate rain forest Temperate forests vary considerably with climate, and those occurring in exceptionally wet conditions are sometimes referred to as temperate rain forests. Areas that have been so termed include the north-west Pacific coast of North America, the windward slopes of the Andes in southern Chile, the southern part of the South Island of New Zealand, and parts of eastern China and Japan. [P.D.M.]

temperature Temperature is a property of a body or region of space which can be sensed subjectively as relative hotness. It is a measure of the average kinetic energy status of the molecules or atoms making up the system. If there is a temperature difference between two bodies then heat will flow between them. Temperature can only be measured by observing the behaviour of another substance which varies predictably with temperature, for example the expansion of mercury or the electromotive force produced by a thermocouple. Temperature scales are based either on arbitrary points of observable reference such as the freezing point or boiling point of water (Celsius scale) or on an absolute scale in which the so-called thermodynamic temperature is a function of the energy possessed by matter (Kelvin scale). [J.B.M.]

terrigenous Referring to OCEAN sediments composed of debris from land erosion. [V.F.]

territoriality Territoriality can be said to arise whenever individuals or groups occupying a suitable habitat (*see* HABITAT SELECTION) space themselves out more than would be expected by chance, spacing being active in some way rather than, say, a passive product of underlying HABITAT STRUCTURE or non-random predation. Such an open-ended definition is robust to the easy exceptions plaguing narrower definitions like 'defended area' or 'exclusive area', and encompasses the full scope of scale and mechanism underlying spacing patterns. Separation may range from a few millimetres between neighbouring limpets on a rock to several kilometres between packs of hunting dogs or herds of antelope. Spacing between neighbours may be maintained in several ways, for instance by aggression, as in many birds and fish, or by vocal or olfactory signals as in some primates and ungulates. The secretion of TOXINS into the soil by some plant species can also be viewed as a means of territorial spacing. Territories may reflect the exclusive use of a given area, or the same area may be shared by different individuals or groups using it at different times. Moreover, they can be a permanent or temporary feature of an individual's life history, often existing on a seasonal basis and being restricted to a proportion of competitive individuals (or groups) within a population. [C.J.B.]

territory size TERRITORIALITY is assumed to reflect adaptive strategies of spacing between organisms. If maintaining a territory is adaptive, we would expect the size of territories to reflect the costs and benefits of their defence, all other things (such as physical and environmental CONSTRAINTS) being equal. This is the principle of 'economic defendability'. [C.J.B.]

tertiary consumer A tertiary consumer is one which occupies the fourth TROPHIC LEVEL in a COMMUNITY. In the grazing FOOD WEB, herbivores are primary consumers, primary carnivores and secondary consumers, and the animals which eat the primary carnivores are the tertiary consumers. [M.H.]

thelytoky The parthenogenetic production of females by dioecious animal taxa. Thelytokous reproduction may be apomictic or automictic, amictic or mictic. *See also* ASEXUAL REPRODUCTION; PARTHENOGENESIS. [M.M.]

theory of natural selection NATURAL SELECTION can be defined as a process occurring under three conditions. If a population has:

1 variation among individuals in some TRAIT;

2 a consistent relationship between that trait and FITNESS, i.e. consistent differences among individuals in fitness;

3 a consistent relationship between parents and offspring for the trait, that is, differences in the trait are inherited;

then:

1 the trait frequency distribution will differ among age classes;

2 the trait distribution will differ between parents and offspring if the population is not in equilibrium.

This description of the theory of natural selection is an updated version of Darwin's theory. [G.D.J.]

thermal pollution Heat release into the environment as a by-product of human activity and having a harmful effect on ecosystems. For example, direct thermal POLLUTION of waters may occur at power-stations where more than 50% of the heat content of the fuel may end up as waste heat that is removed by the cooling waters and is ultimately released into RIVERS. This can raise the temperature of waters to an extent that favours exotic species. It also reduces DISSOLVED OXYGEN causing metabolic STRESS. [P.C.]

thermal stratification An important feature of water bodies that results when differences in TEMPERATURE occur between the surface water layers and deeper layers. As a result of the temperature difference, upper and lower layers differ in density and are resistant to vertical mixing. Such STRATIFICATION is typical in temperate LAKES and COASTAL marine waters during the summer and limits the availability of NUTRIENTS (which would otherwise be upwelled from depth) to primary producers. *See also* LAKE STRATIFICATION. [V.F.]

thermocline That portion of the water column where TEMPERATURE changes most rapidly with each unit change in depth. [V.F.]

thermometry Thermometry is the measurement of TEMPERATURE. This is achieved by using one of several types of sensor:

- mercury-in-glass thermometers;
- thermocouples;
- platinum resistance thermometers;
- thermistors;
- infrared thermometers.

[J.G.]

therophyte A plant life form that survives unfavourable conditions (usually DROUGHT) as a seed. [P.D.M.]

thixotrophic The fluid nature of sands and muds as a result of the application of pressure. [V.F.]

thorn scrub Sparse shrub vegetation characteristic of semi-arid areas which is usually deciduous and richly provided with thorns, thus reducing its palatability to large herbivores. [P.D.M.]

threshold In environmental parlance, the divide between when an effect does and does not take place. [P.C.]

threshold selection Quantitative characters that come in two discrete classes are THRESHOLD characters. An example is the number of digits on the hindpaws of guinea pigs: some guinea pigs have three toes, others have four toes. [G.D.J.]

tidal flat Tidal flats are the gently sloping or effectively level expanses of marine SEDIMENT that may be exposed by low TIDES. [R.S.K.B.]

tides The periodic movement of water resulting from gravitational attraction between the Earth, Sun and Moon, the forces of which act unequally on different parts of the Earth. Spring tides occur near the times of the new and full Moon, when the Sun and Moon lie on a straight line passing through the Earth and their influences are additive. Tidal range is greatest during spring tides. Neap tides occur near the first and last quarters of the Moon, when the Sun and Moon lie at right angles to each other and have directly opposing influences. Tidal range is least during neap tides. A DIURNAL tide has one high water and one low water each tidal day. A semidiurnal tide is a tide with two high waters and two low waters each tidal day, with small inequalities between successive highs and successive lows. A mixed tide is a tide having two high waters and two low waters per tidal day with a marked diurnal inequality. This tide type may also show alternating periods of diurnal and semi-diurnal components. Variability in tides is caused by the configuration of the ocean's bottom TOPOGRAPHY and by seasonal differences in currents. Tidal currents are caused by the alternating horizontal movement of water associated with the rise and fall of the tide. They are important agents of erosion and significantly affect size, sorting and distribution of SEDIMENT over most of the sea floor. [V.F.]

Tilman's model of resource utilization An optimal foraging model (*see* OPTIMAL FORAGING THEORY) for plant GROWTH which predicts that a plant should consume two essential RESOURCES in the proportion in which it is equally limited by both of them. This proportion defines the consumption vector for the two resources and has important implications for species COEXISTENCE. The model applies in principle to any number of resources. [M.J.C.]

timber line The border between the outmost wood stands and the treeless vegetation of the polar regions, mountains and arid areas. [Y.V.]

time Time is a fundamental physical quantity describing the duration of events as a dimension, or the sequence of a number of events. The SI UNIT for time is the second (s) which is defined relative to the frequency of transition between two levels of the ground state of a ^{133}Cs atom. Most other common units of time are defined with reference to the second.

Systems describing time were originally based on the DIURNAL cycle of the Earth's rotation around it's own axis and the seasonal cycle resulting from the orbit of the Earth around the Sun.

True solar time or local apparent time is based upon the apparent motion of the Sun across the sky. Solar noon is defined for any locality as the time when the Sun reaches its highest point or solar angle. This point is known as the local meridian. The period between two meridians is defined as 1 day. The length of the solar day varies by up to 16 minutes during the year because of the position of the Earth in orbit around the Sun. The variation in length of the solar day is described by the equation of time often published as part of ephemeris tables (e.g. Nautical and Astronomical Almanacs). In many biological applications, true solar time must be calculated from local standard time making allowances for longitude, the equation of time and daylight saving. The mean solar day length averaged over a year is used by convention for most purposes. Mean solar time is defined relative to the movement of a hypothetical sun travelling across the sky at the same rate throughout the year. Sidereal time is defined using fixed stars as a reference, with a 'sidereal day' being the period required for the Earth to rotate once relative to a fixed star. The length of a sidereal day is approximately 4 minutes shorter than the mean solar day. Ephemeris time is based upon the annual revolution of the Earth around the Sun, removing the difference associated with the variation in the speed of rotation of the Earth about its axis.

Standard time was developed from mean solar time to avoid complications for travel and communications associated with communities using their own local solar time. The Earth was divided into 24 standard time zones, each separated by 15° longitude. True solar time differs from standard time by 4 minutes for each degree of longitude east or west of the meridian for the time zone. The reference standard time zone is located at the 0° meridian of longitude that passes through the Royal Greenwich Observatory in England. The local mean solar time at the Greenwich meridian is referred to as Greenwich Mean Time (GMT) or Universal Time (UT). Standard time zones are described by their distance east or west of Greenwich but are sometimes modified for political or geographical reasons. Daylight saving is a system where clocks are set 1 or 2 hours forward during the summer months, resulting in an additional period of daylight in the evening.

The GEOLOGICAL TIME-SCALE is a record of the Earth's history defined relative to major geological episodes. Divisions of geological time are based on changes in the FOSSIL RECORD in geological strata. [P.R.V.G.]

time-series analysis The set of statistical techniques used to suggest likely ecological mechanisms that might have generated a particular run of POPULATION DENSITY data (the time series). Generally, the time series is de-trended prior to analysis. [M.J.C.]

tolerance The reduced response to a TOXICANT, BIOCIDE or pollutant as a result of repeated exposure. This may occur within individuals as a result of physiological ACCLIMATION/acclimatization—for example as a result of induction of protective, metabolic systems such as melallothioneins or mixed-function oxidases (MFOs). It may also occur by selection of more tolerant genotypes. [P.C.]

top-down controls Important influence on COMMUNITY structure and possibly function from upper trophic levels. Thus removal of a predator releases pressure on prey which has knock-on effects for its competitors and the species populations that form its food. *See also* HOLISM; SYSTEMS ECOLOGY. [P.C.]

top species The species at the ends of FOOD CHAINS, i.e. the pinnacles of trophic pyramids. Also known as top carnivores or top PREDATORS—they are invariably meat-eaters. *See also* FEEDING TYPES, CLASSIFICATION. [P.C.]

topogenous Used of a MIRE that owes its development to the concentration of water in a region by drainage from a CATCHMENT. By definition, therefore, such a mire must be flow-fed, or RHEOTROPHIC, for example basin fens and some parts of valley mires. [P.D.M.]

topography The surface features of the Earth, including the relief, the terrain, the VEGETATION, the SOILS and all the features within the landscape. It is not synonymous with relief alone. [P.J.C.]

topology The branch of mathematics concerned with properties of objects that are not distorted by stretching, etc. [M.J.C.]

toxic bloom One possible outcome of EUTROPHICATION is the appearance of toxic algae, i.e. those algae, most usually blue-greens, capable of excreting substances that are toxic to wildlife and/or domestic animals and/or humans. These algae can dominate algal communities and become visible as coloration—a BLOOM—within the water. This can occur in both inland and COASTAL waters. A well-known example in recent years has been the blooms of dinoflagellates in the coastal areas of Skagerrak and Kattegat, close to the North Sea, that caused considerable harm to the AQUACULTURE industry, and may, by weakening their immune system, have led to an increased prevalence of disease and death in the seal population in that area. [P.C.]

toxicant Artificially produced substances that have poisonous effects on humans or wildlife. They may be produced as intended poisons, i.e. BIOCIDES, or they may be produced for other reasons but have toxic/ecotoxic effects. *Cf.* TOXINS. [P.C.]

toxins Natural poisons produced by organisms, for example plant poisons, snake and spider venoms, and antibiotics. They are usually produced as intended poisons. *Cf.* TOXICANT. [P.C.]

trade-off The exchange of one advantageous character for another. In the evolutionary process there is always NATURAL SELECTION of organisms that have higher growth rates or fecundities, or lower mortality rates. Trade-offs occur when improvement of one of these life-history characters can only be achieved at the expense of another/others. For instance, increasing fecundity may entail a mortality cost, so that there is a COST OF REPRODUCTION. [R.M.S.]

trait Character, usually presumed to be under genetic control and hence open to SELECTION. It can be 'simple' (e.g. eye colour), part of a complex (e.g. life-history feature) or a complex (e.g. semelparous life history). [P.C.]

trait group A group of features or characters, possessed by an individual, a species or other taxonomic grouping. TRAIT is a term used for a particular PHENOTYPE in human genetics (e.g. sickle-cell trait—the phenotype of the haemoglobin-A/haemoglobin-S heterozygote). [L.M.C.]

trajectory The path in a phase plane graph traced by a series of VECTORS (MATHEMATICAL) over time. The trajectory shows the joint densities of the two species over time and the speed of change. [B.S.]

tramp species Animal (usually bird) species found on most tropical and subtropical oceanic islands. These animals often live in habitats created directly or indirectly by human settlement. *See also* ASSEMBLY RULES; DISTRIBUTION. [M.J.C.]

transect A line along which the DISTRIBUTION of the VEGETATION or fauna of an area is investigated, all plants or animals touching an extended tape being recorded. Transects are frequently used in zoned vegetation. [A.J.W.]

transfer efficiency *See* ECOLOGICAL EFFICIENCY; LINDEMAN EFFICIENCY.

transformation of data There are three reasons why the original observations recorded by an ecologist need to be transformed or converted into more suitable units. These are:

1 for visual convenience in a graph;

2 to convert a curved relationship into a more convenient straight line; and

3 because the original data is unsuitable for PARAMETRIC STATISTICS.

[B.S.]

transgenic organisms Transgenic organisms are organisms which have been modified by the insertion of 'foreign' nucleic acid to allow the expression of a new TRAIT or PHENOTYPE which would not normally be observed for that organism. [J.K.]

transition mire This term is used of RHEOTROPHIC mires in which the flow of water is slow and the mineral nutrient content of the water is poor, leading to the development of an acidophilous vegetation intermediate between poor FEN and OMBROTROPHIC bog. Often such conditions may lead on to the development of true bog vegetation in the course of further PEAT accumulation. [P.D.M.]

transitive competition This term is applied to multispecies interactions when a fixed or stable hierarchy of INTERSPECIFIC COMPETITION is present. It is used to classify the pattern of interaction present between all directly competing species. For example, species A may displace species B which may in turn displace species C, which may *also* be displaced by species A. [R.C.]

transmission rates of parasites This is the rate at which hosts become parasitized, it is also known as the incidence of infection and for macroparasites it is the rate of growth of the PARASITE population. [G.F.M.]

transmission threshold For microparasites the TRANSMISSION RATE OF PARASITES may be dependent on the DENSITY of susceptible individuals (i.e. those able to be infected), and the survival of the PARASITE in the HOST population is dependent on the rate of supply of susceptible hosts (*see* EPIDEMIC and OUTBREAKS). There is a critical density of susceptible hosts (given as the reciprocal of the basic reproduction number) below which either an epidemic cannot occur or the parasite will be eradicated from the population. The object of immunization and vaccination programmes is to keep the density of susceptible hosts below the critical density, and therefore the transmission rate below the threshold.

For macroparasites the transmission threshold can be defined in terms of the probability of a parasite finding a host, and is thus related to the host density. However, the concept is less widely used in this context. [G.F.M.]

transpiration Loss of water from the plant surface by evaporation through the stomata. [J.G.]

transplant experiments The transfer of individuals, or samples of populations or communities, from one place to another. Usually done to investigate the importance of the physical/geographical location in characteristics being observed. [P.C.]

tree line The limit of isolated bigger or smaller FOREST stands and single trees towards treeless vegetation in polar regions, on high mountains and in arid continental regions. [Y.V.]

trophic cascade hypothesis The hypothesis states that nutrient input level sets the potential productivity of lakes and that deviations from the potential are due to food-web effects. Variations in the intensity of predation by piscivorous fish in lakes are transmitted sequentially (or cascade) from predators to prey through each step of the FOOD WEB such that abundance, composition and productivity of the basal TROPHIC LEVEL (the PHYTOPLANKTON) is constrained by activities of the top level of the food web.

In LENTIC systems, piscivores, when present, can determine the size and composition of the planktivorous fish assemblage beneath them in the food web. In turn, large herbivores are preferentially consumed by planktivorous fish. Thus abundant planktivorous fish can shift zooplankton composition towards DOMINANCE by smaller individuals, but when these fish are absent, predation by planktivorous invertebrates and COMPETITION among herbivores shift the zooplankton community towards larger sized species. These larger individuals, especially large cladocerans like *Daphnia*, have a greater impact on phytoplankton. According to the trophic cascade hypothesis, algal biomass and primary production would be less in *Daphnia*-dominated lakes (with piscivores, thus lacking planktivorous fish) than in lakes dominated by smaller zooplankton such as *Bosmina*, small calanoid copepods and rotifers (that lack piscivores and possess large populations of planktivorous fish). [P.S.G.]

trophic classification The majority of freshwater animals are generalists (POLYPHAGOUS) rather than specialists (MONOPHAGOUS or OLIGOPHAGOUS) which has ecological implications regarding COMPETITION and temporal or MICROHABITAT isolation. Using a combination of food categories and feeding mechanisms, aquatic invertebrates can be classified into several functional feeding groups as follows.

1 Shredders (chewers and miners)—feeding on COARSE-PARTICULATE ORGANIC MATTER (CPOM; $>10^3\,\mu m$) such as decomposing allochthonous leaves and the associated bacteria and fungi or living vascular plants (gougers).

2 Collectors—feeding on FINE-PARTICULATE ORGANIC MATTER (FPOM; $<10^3\,\mu m$) and associated bacteria and fungi either suspended in the water (filter- or suspension-feeders) or deposited on the substrate (collector–gatherers).

3 Scrapers (grazers)—herbivores feeding on algae and associated material (PERIPHYTON).

4 Piercers—of living macrophyte cells or filamentous algae.

5 Predators—including those which ingest whole prey or parts (swallowers) and those which pierce the prey and extract cells and tissue fluids (piercers).

6 Parasites—internal and external parasites feeding on living animal tissue.

This classification distinguishes taxa performing different functions within LENTIC and LOTIC ecosystems and reflects CONVERGENT and PARALLEL EVOLUTION. *See also* FEEDING TYPES, CLASSIFICATION. [R.W.D.]

trophic continuum This concept provided an alternative model to the well-established Lindeman TROPHIC LEVEL concept of ECOSYSTEM structure. Often species cannot readily be assigned or confined to a single trophic level and their FEEDING activity spans several levels in a traditionally defined FOOD WEB. Such species have often been described as occupying a trophic CONTINUUM. [P.S.G.]

trophic group amensalism hypothesis This hypothesis was developed to explain patterns of low DIVERSITY, high DENSITY and DOMINANCE by certain trophic FEEDING groups in muddy marine benthic habitats. There often appears to be a negative relationship between deposit-feeding invertebrates (especially surface deposit-feeders) and filter-feeding (or suspension-feeding) invertebrates, the former dominating on muddy substrates. The hypothesis assumes that feeding activities of the deposit-feeders resuspends SEDIMENT, thus providing biologically induced instability in the substrate. This inhibits filter-feeders by clogging filtering mechanisms and makes the substrate unsuitable for larval SETTLEMENT and maintenance of filter-feeders, i.e. there is a form of ammensal interaction between the deposit-feeders and the filter-feeders (the latter trophic group being negatively affected by the interaction and the former group unaffected). The DISTURBANCE of the substrate is, in turn, predicted to lead to low-diversity communities dominated by a high ABUNDANCE of deposit-feeders. [P.S.G.]

trophic level A trophic level is a position in a FOOD CHAIN determined by the number of energy transfer steps to that level. [M.H.]

trophic ratios The trophic or PREDATOR:prey ratio is the ratio of the number of species of predators to the number of species of prey within the COMMUNITY. This ratio appears to be relatively constant for a wide variety of communities and independent of SPECIES RICHNESS. Earliest assessments of community food webs broadly defined 'kinds' of predators and prey (trophospecies) and found a predator:prey ratio slightly greater than 1. Similarly, the proportion of all species that are 'top' (predators with no predators), 'intermediate' (both predators and prey) and 'basal' (prey with no prey) trophospecies also appears to be relatively constant, irrespective of the total number of species in the FOOD WEB (average proportions 0.285, 0.525 and 0.189, respectively). [P.S.G.]

trophic structure The trophic structure of a COMMUNITY is a way of describing its structure by reference not to its taxonomic species composition but by reference to the FEEDING relationships of the organisms within it. [M.H.]

tropical montane forest Above about 1200–1500 m the composition of the tropical FOREST changes. Various kinds of tropical montane forest can be recognized. The most common kind is broad-leaved. Microphyllous or needle-leaved coniferous or ericaceous forest is another kind, and in some places bamboo forest is present. For example, in Malaysia the lowland dipterocarp forest is replaced by a broad-leaved montane forest in which species of oak and laurel are common. At about 1800 m the oak–laurel forest is replaced by a montane ericaceous forest. There is a decline in the height of the CANOPY from about 30–40 m in the dipterocarp forest to 20 m in the oak–laurel forest and 15 m in the ericaceous forest. In the montane forest, the canopy is relatively uniform, with few emergent trees. In many parts of the tropics, where it is particularly wet and light levels are low, there is a CLOUD FOREST or, at higher altitude, an elfin forest. [M.I.]

tropical rain forest The term 'tropical rain forest' was coined in the 1890s. The lowland equatorial forests are located mainly in South America (about 50% of the world's tropical RAIN FOREST are found in this region), south-east Asia (about 30%) and central Africa (20%). The vegetation is typically dominated by EVERGREEN trees whose buds lack scales or any protection against DROUGHT or cold. Mean annual precipitation usually exceeds 2000 mm and may be greater than 4000 mm. Temperature shows little seasonal fluctuation and usually lies around 26–30°C. [P.D.M.]

tropical seasonal forest Rainfall is relatively uniformly distributed throughout the year in many equatorial regions, as in south-east Asia generally, and also in much of Brazil. But other areas have an uneven rainfall pattern, such as southern Nigeria, coastal Zaire and parts of Central America. The incidence of distinct dry and wet seasons, particularly as one moves away from the equatorial regions and into the monsoon zone (*see* MONSOON FOREST), leads to new selective pressures on plant and animal species. Seasonal tropical forests may still be dominated by EVERGREEN species of tree, or by species that shed all or a large proportion of their leaves as the DRY SEASON develops. Buds may also be protected by coverings in such species. Climbers and epiphytes are still present, but are less abundant than in the more open of the equatorial rain forests and become less frequent with the increasing severity of the dry season. [P.D.M.]

troposphere The weather-filled part of the ATMOSPHERE, bounded below by the Earth's surface and above by the tropopause (10–15 km above sea level), and containing 70–90% of its mass (the larger in low latitudes) and almost all of its water vapour and cloud. It is continually disturbed by a wide variety of types of weather system, and has typically steep TEMPERATURE lapse rates associated with widespread convection.

The lowest few hundred metres (the atmospheric BOUNDARY LAYER) is directly affected by the underlying surface through friction, surface shape, heating and cooling. [J.F.R.M.]

tropospheric chemistry The chemical processes occurring in the TROPOSPHERE by which trace substances released into the ATMOSPHERE from natural and anthropogenic sources are oxidized. The reactions form stable compounds, such as carbon dioxide (CO_2) and water, or more soluble compounds, such as peroxides or acids, which are rained out or deposited from the atmosphere. [P.B.]

tundra In origin from Finnish or Lappish *tunturi* meaning a treeless hill which has been absorbed via Russian to describe the dwarf shrub, herb and moss vegetation that exists in polar regions too cold to support tree growth. The treeless nature is due in part to the presence of permafrost sufficiently near the soil surface which restricts the depth of soil available for tree root development and necessary to support full-grown trees. Exceptions to the exclusion of trees from permafrost zones are the black spruce of North America (*Picea mariana*) and the dahurian larch of north-eastern Siberia (*Larix dahurica*). Both these species are able to extend shallow root systems over a soil ice horizon and thus support substantial trees in permafrost areas. [R.M.M.C.]

turbidity Reduced transparency of either ATMOSPHERE (caused by absorption and scattering of radiation by solid or liquid particles in suspension) or liquid (caused by scattering of light by suspended particles). [P.C.]

turbulence The chaotic and seemingly random motion of fluid parcels is known as turbulence. [D.D.B.]

turbulent flow Fluid flow that exhibits chaotic and unstable behaviour is turbulent. [D.D.B.]

turnover Turnover in ecology broadly refers to the flux of energy and matter (i.e. species, individuals, biomass or nutrients) through an ECOSYSTEM. Species turnover is widely used to describe the replacement of one species by another in an ASSEMBLAGE. The phenomenon of species turnover may be considered in both space and time. Spatially, the replacement of species along ecological and geographical gradients is sometimes termed beta diversity (*see* DIVERSITY, ALPHA, BETA AND GAMMA). Examples of assemblages with both high and low levels of species turnover have been documented. *See also* BIOGEOCHEMICAL CYCLE; LAKE STRATIFICATION; MACARTHUR–WILSON MODEL OF ISLAND BIOGEOGRAPHY. [N.M. & S.G.C.]

two-tailed test In the NORMAL DISTRIBUTION 95% of observations are contained within $\mu \pm 1.96\sigma$ and the remaining 5% of observations are equally divided between the two tails of the distribution, 2.5% in each tail. When we carry out a *z(d)* TEST to examine the difference between two means, and we have no a priori reason to expect a value of the test statistic above or below $z = 0$ (either MEAN could be larger), we use the 5% level of significance that corresponds to $z = 1.96$. This is a two-tailed test and corresponds to a NULL HYPOTHESIS of H_0: $\bar{x}_1 = \bar{x}_2$ and an alternative hypothesis of H_1: $\bar{x}_1 \neq \bar{x}_2$. Any statistical test that uses the PROBABILITY in both tails of a distribution is therefore a two-tailed test. The alternative is to use the value of the test statistic that cuts off 5% in just one tail. For the normal distribution this would correspond to $z = 1.65$. Using this value as the 5% significance point would imply a ONE-TAILED TEST. [B.S.]

type 1 error The rejection of a NULL HYPOTHESIS when it is actually true. For example, if we use a *t* TEST to examine the difference between the means of two samples (null hypothesis that they come from populations with the same MEAN: $\mu_1 = \mu_2$) and conclude that they are significantly different when they are not (i.e. conclude that $\mu_1 \neq \mu_2$ when in fact $\mu_1 = \mu_2$). The PROBABILITY of making such a type 1 error, with a *t* test, would be $\alpha = 0.05$ if we accept the 5% level of significance. [B.S.]

type 2 error The acceptance of a NULL HYPOTHESIS when it is actually false. For example, if we use a *t* TEST to examine the difference between the means of two samples (null hypothesis that they come from populations with the same MEAN: $\mu_1 = \mu_2$) and conclude that they are not significantly different when they are (i.e. conclude that $\mu_1 = \mu_2$ when in fact $\mu_1 \neq \mu_2$).

Scientists are often more worried about type 1 than type 2 errors; making false claims about treatments. However, a situation where there might be a concern about type 2 errors is in accepting no difference between polluted and non-polluted sites when in fact they exist. [B.S. & P.C.]

U

ultraviolet radiation Electromagnetic radiation in the wavelength interval between X-rays and visible light, i.e. from *c.* 100 nm to 400 nm (these limits are not universally agreed upon; some scientists set the lower limit at 5 nm or even lower, or the upper limit at 380 nm). Ultraviolet (UV) radiation was discovered in 1801 by the German, Johann Wilhelm Ritter, who found that the radiation outside the violet end of the visible solar spectrum could decompose silver chloride (AgCl).

UV radiation has effects on other biological systems, such as PHYTOPLANKTON and terrestrial plant communities. Stratospheric ozone depletion has caused concern about the consequences that increases in such effects may have on a global scale. Much attention has been devoted to UV-B radiation in recent years because its intensity at ground level and in the sea and other waters increases with depletion of the atmospheric ozone. There has been a trend of decreasing stratospheric ozone over the past couple of decades (except near the Equator), almost certainly due to anthropogenic POLLUTION of the ATMOSPHERE by CFCs and nitrogen oxides (NO_x). Although this has resulted in international regulations to protect the atmosphere, return to normal conditions will take many decades even if agreements are adhered to. Until now, the effects on UV-B of the depletion of stratospheric ozone have been partly compensated for by other pollutions causing increases of tropospheric (low-level) ozone and haze, and it has not been firmly established that present-day UV-B levels are unnaturally high, except under the ANTARCTIC OZONE HOLE. [L.O.B.]

umbrella species Species whose occupancy area (HOME RANGE) is large enough and whose habitat requirements are wide enough that if they are adequately protected, other species will automatically be protected as well. [P.C.]

uncertainty Lack of understanding or control of situations due to: (i) ignorance; (ii) poor EXPERIMENTAL DESIGN and/or sloppy observation; and/or (iii) stochasticity in the process being studied. [P.C.]

undercompensation Occurs when density-dependent MORTALITY is weak, such that restoration of a population equilibrium following disturbance takes more than one generation. Density-dependent mortality is detected by plotting $k = \log_{10}$(initial density/final density) against $\log_{10}$(initial density); undercompensation occurs if the slope $b < 1$. [R.H.S.]

unit Quantity or dimension adopted as a standard for measurement. The scientific community has adopted SI UNITS (Système International d'Unités) as the standard. [J.G.]

unit leaf rate (ULR) A measure of the net rate of carbon fixation by plants, expressed as the amount of dry matter (e.g. grams) fixed per unit time (e.g. per day) per unit leaf area (e.g. per square metre). Units are $g\,m^{-2}\,time^{-1}$. [M.J.C.]

unitary organism A unitary organism has a fixed morphological form which persists throughout its adult life. Any changes in form are almost always associated with transitions between juvenile and adult stages as is the case in many insect orders. The pattern of development is fixed and is almost totally independent from the surrounding environment. For instance, the morphological form of mammals is fixed at an early embryonic stage and the basic plan is then retained until death. *Cf.* MODULAR ORGANISM. [R.C.]

units of selection Any biological entity that replicates itself and retains its identity from generation to generation is a unit of selection. A unit of selection is a selectable entity. Genetic units of selection include individual genes, chromosomes, genotypes, populations and species. In clonal, or asexually reproducing species, the entire GENOTYPE is reproduced in the offspring, and the individual genotype can be considered to be a unit of selection. In sexually reproducing species, individual organisms consist of unique collections of genes that are not transmitted as a unit to the next generation. However, the single genes upon which selection acts do not replicate independently, but rather depend on the reproduction of the entire organism in which they are carried. Thus the FITNESS of the individual genes is determined by the whole of the genotype and the fitness of an individual gene may, in fact, be heavily dependent on its interactions with other genes carried by the organism. A meaningful measure is a gene's (or allele's) average fitness, which is the average fitness of all of the individual's carrying it (each with their own unique genotype). Populations and species may also act as units of selection in that some will originate or go extinct more rapidly than others, resulting in a shift in the mean PHENOTYPE of the entire group. [V.F.]

unmatched observations Observations made on different experimental or observed units. [B.S.]

unstable equilibrium An equilibrium is unstable if a perturbation away from equilibrium tends to increase with time, rather than diminishing. It is sometimes possible that a small perturbation will diminish whilst a large one will increase, in which case the equilibrium is locally stable but globally unstable. Systems with unstable equilibria often show cycles or CHAOS, but may also become extinct or grow without limit. [S.N.W.]

upstream–downstream comparison Comparison of the BIOTA above and below an effluent stream emptying into a stream or river. Upstream is used as a comparator for deviations from normality downstream, i.e. it is presumed that the downstream system would be like the upstream in the absence of STRESS—but this need not be so. *Cf.* BACI. [P.C.]

upwelling Process by which deep, cold, nutrient-laden water is brought to the surface, usually by diverging equatorial currents or COASTAL currents that pull water away from the coast. Upwelling is especially conspicuous on the eastern sides of ocean basins, and such coastal upwelling regions are the most productive areas in the ocean. The nutrient-rich waters generally promote PHYTOPLANKTON growth which in turn supports rich ZOOPLANKTON and fish populations. Short trophic webs in upwelling areas mean more efficient energy transfers relative to the open ocean where trophic webs are longer. Upwelling areas represent about 0.1% of the ocean surface area but 50% of the the world's fishery. [V.F.]

V

vacant niches Niches representing RESOURCES and/or conditions that are not utilized. [J.G.S.]

vagility Ability to disperse. [P.C.]

variability Changes in ecological properties and processes through space and time. Also refers to changes in results from experiments carried out simultaneously (replicability), or at different times in the same laboratory (REPEATABILITY), or at different laboratories (REPRODUCIBILITY). It can be measured as a COEFFICIENT OF VARIATION. *See also* POPULATION DYNAMICS. [P.C.]

vector, mathematical A quantity having both size and direction as opposed to a scalar quantity that has size only. Ordinary numbers (which have only size) are therefore scalars, but a force (which has a size and is applied in a particular direction) is a vector. In ecology they occur in phase plane graphs/diagrams, in which the dynamics of two-species systems (COMPETITION or predator–prey) are depicted. Each point on the graph represents a joint DENSITY of the two species. To each point in this graph we can attach an arrow, or vector, whose length and direction will indicate the dynamics of the system (*see* LOTKA–VOLTERRA MODEL). [B.S.]

vector, parasites Parasites must transmit between definitive hosts (*see* SECONDARY HOST), and if a second species acts to convey the PARASITE between hosts, it is known as a vector species. Parasites may have more or less specificity to different vectors. The vector species may suffer reduced FITNESS when infected by the parasite (but need not). [G.F.M.]

vegetation The sum total of plants in the broadest sense (tracheophytes, bryophytes and algae) at a site, by convention including certain symbiotic fungi (mutualistic mycorrhizae and mycorrhizae parasitized by orchids and a few other specialized plants) but not the fungi parasitic or saprophytic on the plants. [P.J.G.]

vegetative reproduction Reproduction by somatic division. In plants this widespread method of multiplication may involve, for example, rhizomes, stolons (runners), tubers, bulbs, corms, suckers, gemmae or turions (swollen detached buds of aquatics). Many of these structures also serve as perennating organs. [R.N.H. & A.J.W.]

veld The temperate GRASSLANDS of the eastern area of the southern part of Africa. Scattered trees and bushes may be present, particularly at lower altitudes. [P.D.M.]

vernal pools Seasonally wet, shallow pools, commonly found in the Central Valley and surrounding lower foothill regions of the coastal and Sierra Nevada ranges in California. [S.K.J.]

vertical migration Movements of PELAGIC organisms on a daily or seasonal cycle that may be associated with FEEDING or breeding. The typical DIEL pattern is movement upward during the night and movement downward during the day. The distance travelled is species dependent, but even small copepods may travel several hundred metres twice in a day. Various explanations have been proposed to explain this behaviour, including escape from visual predators, conservation of energy by time spent in deeper colder water, and exposure to new feeding areas. [V.F.]

vicariance biogeography A type of BIOGEOGRAPHY which is concerned with discovering the commonality of observed distributional patterns of organisms shown by evolutionarily unrelated taxa, which suggests a common and simultaneous process. [A.H.]

vicarious distribution This is a type of DISTRIBUTION produced after BIOTA are split up (vicariated) by the emergence of barriers through, for example, CONTINENTAL DRIFT, which separate parts of a once continuous biota and alter the proximity of populations or organisms irrespective of their own movements. [A.H.]

vigilance Prey can avoid predators by increasing the amount of time and energy they spend trying to sense nearby predators. Once a PREDATOR is detected, suitable measures can be taken, such as flight, defence or enhanced vigilance. There may well be a TRADE-OFF between predation risk and foraging success mediated by the degree of vigilance. Vigilance is best understood in vertebrates, particularly birds. [D.A.B.]

virulence That aspect of PARASITES which measures the harm done to the HOST, although usually only used for microparasites. Usually meaning the case MORTALITY RATE of a parasite (i.e. the risk that an individual host will die when infected), although, generally, the harm may be considered as a combination of MORBIDITY (disease) and mortality (sometimes referred to as pathogenicity of the parasite).

It is also used to describe the effect of the parasite on the host population level, in which case it also includes the aspects of the parasite's ability to infect hosts, such that a parasite with a high basic reproduction number (*see* TRANSMISSION RATES OF PARASITES) may be more virulent. [G.F.M.]

viviparity The production of liveborn young. [D.N.R.]

von Bertalanffy equation A simple equation used to describe somatic growth. If W represents body mass and t is age:

$$dW/dt = aW^m - bW$$

where a, b and m are constants. The first term on the right-hand side represents the rate of anabolism—the synthesis of tissues—and the second term represents the rate of their breakdown. The equation is usually used in an integrated form for the case $m = \frac{2}{3}$, when:

$$L(t) = L_\infty(1 - ce^{-kt})$$

where $L(t)$ represents body length at age t, and L_∞, c and k are constants. L_∞ represents asymptotic body length, c depends on initial body length, and k is sometimes called the 'growth coefficient'. k is low if the organism takes a long time to approach its final size, and high if it approaches its final size rapidly. [R.M.S.]

W

water (hydrological) cycle The natural cycle of evaporation, condensation and rainfall. On the Earth's surface, 97% of the water is in the oceans, and most of the remainder is in the ice caps. Only a tiny fraction of the total is in the ATMOSPHERE at any time (Table W1). Each day about 1200 km³ of water evaporates from the oceans and 190 km³ is lost from the terrestrial surface by evapotranspiration. Around 1000 km³ falls on the SEA as rain and other precipitation, and 300 km³ falls over the land. A flow of some 110 km³ is returned to the sea via RIVERS, groundwater and meltwater.

Table W1 Global water reservoirs and fluxes. (After Ward, R.C. & Robinson, M. (1990) *Principles of Hydrology*. McGraw-Hill, Maidenhead. Reproduced with the kind permission of McGraw-Hill Publishing Company.)

	Values ($km^3 \times 10^3$)	Percentage of reservoir
Global reservoir		
Ocean	1 350 000	97.403
Atmosphere	13	0.001
Land	35 978	2.596
Terrestrial reservoir		
Rivers	2	0.006
Freshwater lakes	100	0.278
Inland seas	105	0.292
Soil water	70	0.195
Groundwater	8200	22.792
Ice caps/glaciers	27 500	76.436
Biota	1	0.003
Annual flux		
Evaporation		
Ocean	425.0	
Land	71.0	
Total	496.0	
Precipitation		
Ocean	385.0	
Land	111.0	
Total	496.0	
Run-off to oceans		
Rivers	27.0	
Groundwater	12.0	
Glacial meltwater	2.5	
Total	41.5	

In practice there are numerous short circuits to this cycle. Water, having fallen on to the Earth's surface, may travel back into the atmosphere through any of a number of different pathways. On land, some of the precipitation will be intercepted by the vegetation and evaporated directly back into the atmosphere. Unless the SOIL is saturated, when overland flow may be generated, the precipitation reaching the ground enters the soil, and is either taken up by plants and evaporated as TRANSPIRATION, or drains through the soil to reach rivers and then the oceans, either directly or via groundwater.

Average residence times for the different phases of the hydrological cycle vary from a few days, for the atmospheric phase, to sometimes more than hundreds of years for groundwater. The size of the global water reservoirs and fluxes is given in Table W1. [J.H.C.G. & J.G.]

water quality objective/standard (WQO/WQS) In UK law, WQO is the target for water quality in a stretch of river, for example to supply fisheries, abstraction for drinking, etc. WQS is a quantitative concentration below which the defined environmental/ecological objectives are likely to be achieved. Elsewhere there is a tendency to use WQO and WQS interchangeably to refer to standards. [P.C.]

water table The water table is the upper envelope of the subsurface zone which is saturated with water, it separates groundwater from the SOIL MOISTURE in the unsaturated zone above. Water tables are dynamic, having depths which are likely to have a seasonal variation in response to the higher rate of groundwater recharge during the winter, when rainfall may be higher and evaporation is less. [J.H.C.G.]

watershed Drainage basin. *See also* CATCHMENT. [P.C.]

weed (i) A plant where I don't want it; (ii) a plant whose virtues have yet to be discovered; (iii) a plant in the wrong place; (iv) a plant which is the target of HERBICIDE application; (v) a plant for whose removal or destruction someone is willing to pay money or expend effort. [M.J.C.]

Wentworth scale A geometric grading scheme for expressing sedimentary grain diameters. According to the phi (ϕ) scale, grain diameter is expressed as its negative logarithm to base 2, thus permitting percentage differences between small particles to be conveniently compared to percentage differences between particles of very large size. [V.F.]

wetlands A broad term for a wide range of habitats characterized by their abundant water supply. [P.D.M]

wildlife In principle, wildlife refers to all the indigenous flora and fauna in nature. In practice, its use often intends only animal life. [P.C.]

wildlife management Actions aimed at conserving indigenous WILDLIFE. [P.C.]

wind The ATMOSPHERE in motion. [J.G.]

wind power The use of WIND vanes to harness energy from the wind in driving electrical turbines. [P.C.]

within-lake habitat classification In LAKES the substrate from the shore down to the deepest parts of the water exhibit a series of zones.

1 Epilittoral—entirely above the water, uninfluenced even by spray.

2 Supralittoral—entirely above the water but affected by the spray.

3 Littoral—from the edge of the water to the point of light penetration where PRIMARY PRODUCTIVITY equals RESPIRATION (compensation point). The littoral is subdivided into:

(a) eulittoral—between the highest and lowest seasonal levels;

(b) upper infralittoral—with emergent rooted vegetation (rhizobenthos);

(c) middle infralittoral—with floating leaves on the rooted vegetation;

(d) lower infralittoral—with submerged rooted vegetation.

Typically, the rhizobenthos forms concentric rings within the LITTORAL ZONE with one group replacing the next as the water depth changes. In the littoral zone the primary producers are of two types: rhizobenthic and haptobenthic, made up of aquatic macrophytic spermatophytes and phytobenthic algae.

4 The region in the deeper waters which has no rooted vegetation is the profundal where respiration exceeds PRODUCTION, while the transition zone between the littoral and profundal zones is the littoriprofundal.

5 The open water zone is termed the limnetic (or PELAGIC) zone and is equivalent to the EUPHOTIC ZONE, where respiration equals PHOTOSYNTHESIS and the compensation point is usually approximated as 1% of full sunlight.

[R.W.D.]

within-stream/river habitat classification There are two major HABITAT types:

1 in riffles (or rapids) the water is shallow with water velocities high enough to keep the bottom clear of SILT and mud;

2 in pools the water is deeper with velocities low enough that silt and other loose materials can settle on the bottom.

The hyporheos consists of the interstices between the particles composing the substrate forming a middle zone between the surface waters of the stream and the groundwater (aquifer) below.

The hyporheic fauna can be divided into: (i) occasional—consisting of benthic species present on a temporary basis; and (ii) permanent—consisting of species found only in this habitat. It has been suggested that the hyporheos acts as a refuge to the occasional members, by acting as a safety zone during periods of STRESS, such as DROUGHT or very high discharge, and from which they can return to the BENTHOS when recolonization is possible. The hyporheos itself is recolonized mainly from upstream sources, followed by downstream sources and the deeper hyporheos.

There is also a pleuston community associated with the water–air interface of streams but this is never as well developed as in LAKES. [R.W.D.]

woodland Tree-covered land and associated habitats in a matrix of trees. Now usually associated in Britain with small woods consisting mainly of native trees, and with traditional or other non-intensive forms of management in which timber-growing is only one of many objects of management (*cf.* FOREST). In other English-speaking countries, woodland labels open-canopied forest in Australia, but is uncertain in its connotations in North America. [G.F.P.]

X

xenobiotic A foreign substance in a biological system, i.e. a substance not naturally produced and not normally a component of a specified biological system. Usually applied to manufactured chemicals. [P.C.]

xerophile A plant of warm, dry habitats, tolerant of DROUGHT and thriving in dry areas, including deserts and semi-deserts. [R.M.M.C.]

xerophyte A plant with specific adaptations for surviving prolonged periods of DROUGHT. [R.M.M.C.]

xerosere A general term used to describe any SUCCESSION on dry land surface, for example rock (lithosere), sand (PSAMMOSERE) or shingle. [B.W.F.]